Wattis Symposium Series in Anthropology

CALIFORNIA ACADEMY OF SCIENCES
Memoir 21

Mrs. Phyllis Wattis
May, 1996

Contemporary Issues in Human Evolution

Editors

W. Eric Meikle
Institute of Human Origins, Berkeley, California
F. Clark Howell
University of California, Berkeley, California
Nina G. Jablonski
California Academy of Sciences, San Francisco, California

Wattis Symposium Series in Anthropology

CALIFORNIA ACADEMY OF SCIENCES
Memoir 21

San Francisco, California
December 2, 1996

Contemporary Issues
in Human Evolution

TABLE OF CONTENTS

Page

Portrait of Mrs. Phyllis Wattis . . . . Frontispiece

Preface, by *Nina G. Jablonski* vii

Thoughts on the Study and Interpretation of the
Human Fossil Record, by *F. Clark Howell* 1

Paleoanthropology and Preconception, by *Ian Tattersall* 47

Grades and Clades: A Paleontological Perspective
on Phylogenetic Issues, by *Pascal Tassy* 55

Homoplasy, Clades, and Hominid Phylogeny,
by *Henry M. McHenry* 77

The Genus *Paranthropus*: What's in a Name?,
by *Ronald J. Clarke* 93

Origin and Evolution of the Genus *Homo*,
by *Bernard Wood* 105

Current Issues in Modern Human Origins,
by *Christopher B. Stringer* 116

Behavior and Human Evolution, by *Alison S. Brooks* 135

Molecular Anthropology in Retrospect and Prospect,
by *Jonathan Marks* 167

INDEX 187

Preface

This volume represents the edited proceeding of the First Paul L. and Phyllis Wattis Foundation Endowment Symposium, which was held at the California Academy of Sciences on February 6, 1993. It represents the precious first child of two very important parents. Its father was F. Clark Howell, world-renowned paleoanthropologist and Fellow of the California Academy of Sciences. It was Clark who, with the help of dedicated assistants from the Anthropology Department at the Academy, developed the theme of the Symposium and brought the key players together. His efforts were only possible, however, with the blessing, inspiration and financial support of the volume's mother, Mrs. Phyllis Wattis. Phyllis Wattis's passionate devotion to the Academy, and especially to its anthropology programs, has been matched by her extraordinary philanthropy. Her own keen interests in anthropology and archaeology, combined with her commitment to public education, have resulted in the development of an extensive series of educational programs at the Academy that includes the Wattis Foundation Symposia.

The Wattis Symposia, born of Mrs. Wattis's interests in anthropology and prehistory, are aimed at bringing "state of the art" presentations on a seminal topic in anthropology to the general public. The presentations are then echoed in published form for distribution to an even wider public. The volumes of the Wattis Symposium Series in Anthropology, the first of which you are now reading, take their place alongside other numbered Memoirs of the Academy.

The first Wattis Symposium and its associated publication also owe their existence to several other energetic and hard-working individuals. The former Irvine Chair of Anthropology, Dr. Linda Cordell, in collaboration with Deborah Stratmann, also of the Academy's Anthropology Department, worked with Clark Howell to translate the idea of the symposium into practice. Their dedication and success set the tone for the entire series. This first volume was produced largely through the tireless efforts of Dr. Alan Leviton, Director of Scientific Publications for the Academy, who crafted the edited text into a handsome book.

As befits the nature of anthropology today, the Wattis Symposia are envisioned as being explicitly multidisciplinary. The theme of the first symposium, human origins, brought together widely recognized scholars from the fields of paleoanthropology, vertebrate paleontology, archaeology and molecular anthropology. Future symposia will do likewise for different subjects and will, to the greatest extent possible, integrate scholarship from the four fields which constitute anthropology in its broadest sense, namely, cultural anthropology, physical anthropology, archaeology and anthropological linguistics.

It is has been a pleasure to have been involved in the production of this first symposium volume, which, we hope, will stand as a lasting tribute to the vision, generosity and seemingly boundless enthusiasm of Phyllis Wattis.

Nina G. Jablonski
Irvine Chair of Anthropology
California Academy of Sciences
June 14, 1996

Thoughts on the Study and Interpretation of the Human Fossil Record

F. Clark Howell
Laboratory for Human Evolutionary Studies/Museum of Vertebrate Zoology
University of California
Berkeley, CA 94720

Human evolutionary studies are flourishing as evidenced here by a diverse set of symposium contributions. This endeavor has acquired a subject matter for its focus and an ever enhanced armamentarium of perspectives and methodologies drawn largely from the natural and physical sciences. Thus, it is fundamentally an aspect of evolutionary biology, most broadly conceived. As no one can master more than a few such procedures and participate actively and intimately in more than a limited spectrum of the overall endeavor, it has come to be greatly enhanced by accretion of participant fellow scientists, of many callings. These have increasingly been incorporated within the endeavor, rather than largely serving it from without, and in so doing are accountable for its attendant transformation. As a consequence it has manifestly gained in breadth as also in depth, and garnered a scientific accountability, largely absent heretofore. Much of this development is often and variously reflected in these contributions. In spite of the accelerated pace of discovery and of manifold field, laboratory, and comparative researches, the endeavor has been bedeviled by controversy. One may justifiably question the extent to which normal science is practiced and a common paradigm, as a system for examining the world, recognized and pursued. I discuss here briefly three areas in which re-examination and redirection of perspective and procedure are sorely warranted. Such are (a) the source, circumstances, and timing of the initial hominin dispersal into Eurasia; (b) the treatment of hominin fossil representatives as paleodeme (p-deme) samples, both independent of and preparatory to the delineating and evaluation of clades (lineages); (c) some evidence relevant to considerations of *Homo sapiens* origins and differentiation, the overall impact of which is rejection and dismissal of hypothetical construals of so-called multiregional continuity models and their correlatives.

This volume is the outcome of a day-long symposium, held at the California Academy of Sciences (Golden Gate Park, San Francisco) on February 6, 1993. Its purpose was to expose to a lay audience basic theoretical, conceptual, and methodological issues about the way we approach and thus hope to understand the evolution of humankind. The symposium was planned by the Academy's Anthropology De-

Contemporary Issues in Human Evolution
Editors, W.E. Meikle, F.C. Howell, & N.G. Jablonski

partment, in consultation with several external advisors. It was realized through the generosity of the Paul L. and Phyllis Wattis Foundation, another example of its continued concern for and support of the CAS over many years. For myself, and for the other nine symposium participants, I wish here to record our great appreciation to the Foundation, and particularly to Phyllis Wattis herself, for the opportunity afforded us on this occasion. One participant was unable to provide a manuscript. Prof. H. M. McHenry was subsequently asked to fill this gap in the proceedings; he generously agreed to do so, and to him the editors offer their warmest appreciation.

I played a small part in the initial planning for the symposium, for two reasons. Firstly, I greatly appreciate and value this Academy and the role it plays both in fundamental research and in broad dissemination of science knowledge. It is a most important institution within the state of California, and particularly so in the Bay Area. And, secondly, because human evolutionary studies center on a search to elucidate our own history, in all its aspects, within the broader context of the emergence, diversification, and (frequently) extinction of life forms on this planet. Thus, it is deservedly every man's study and concern, rather than the preoccupation of a privileged few devotees. Occasions such as this do afford specialists the useful opportunity not only to interact with one another, as often transpires at scientific assemblies, but especially to share with a broader, non-specialist audience their particular concerns and involvement in scientific problems of broad, general interest. This is only as it should be, and certainly human evolutionary studies deservedly have a very substantial following.

There are many facets to human evolutionary studies, only some of which are exemplified in these symposium contributions. These include the higher primate perspective from primatology, the fossil and archeological records and their contexts from paleoanthropology, and the nature and extent of modern human structure and individual and population variability from human biology and molecular biology. The last is among the less well represented here (except, in part, in the contribution by Jon Marks). In each of these major foci there have been major developments, rapid advancements, and shifting perspectives as a consequence of accelerating research programmes, development and application of new methods and analytical proce-dures, and new or different frameworks afforded by altered theoretical perspectives from various affiliated disciplines. Thus, it is oft times difficult to affirm a consensus view in the light of such substantive enhancement of data bases and rapid shifts in approaches to, and analysis of, such expanded resources. Nonetheless, the major outlines and scope of an evolutionary science of paleoanthropology has emerged within the past two decades. Such an endeavor had its roots back at least to the turn of the century, and a long and sometimes painful gestation in the course of the subsequent fifty years. The impact of the maturation and crystallization of modern evolutionary biology, exemplified in the well-known modern synthetic perspective, was somewhat delayed in human evolutionary studies relative to other endeavors in paleobiology. Population genetics, whose application to human populations was initiated some fifty years ago (blood groups), became an integral part of human population biology only some four decades ago (with subsequent expansion of studies of serum proteins and HLA). The explosive effects of molecular biology, as initially applied to primate relationships (protein sequences) are some thirty years old, and as applied more directly to human evolution (analysis of DNA variation) are mostly less than half as old. However, the latter has strong evolutionary and phylogenetic components such that it has tended to enhance and strengthen bridges between human evolutionary biology and human paleobiology (as paleoanthropology). It may well be that some preliminary results of molecular biological investigations of human populations have been initially enthusiastically over-interpreted; nonetheless, it can scarcely be denied that the impact of such researches has been very substantial and that ultimately the phylogenetic and biogeographic significance of such investigations will prove to be indeed far-reaching.

There is no encompassing theory of hominid evolution. (That is, if by theory is meant an argument invoked or constructed to explain why the world is as it appears to be.) Alas, there never really has been despite claims to the contrary drawn from the recitation and some critical analysis of so-called classic texts. For the most part, as evidenced for example in Bowler's (1986) *Theories of Human Evolution*, such are largely speculative excursions and, frequently, comprise inferences in respect to humanity's place in nature, and varied inexplicated accounts of sequence(s) of events inevitably leading to modern humanity. These scenarios or narrative treatments, as Landau (1991) has shown (in *Narratives of Human Evolution*), may often have a common focus on a set of components, entailing transformational events in respect to habitats, lifeways, dietary adaptations, or behavioral potentialities (often in relation to central nervous system structure). These have been recognized as variously 'fossil – free' and, one should add, 'context-poor'. There is a dearth of development of axioms, of theory construction, of hypothesis formulation and testing, etc., in respect to scientific methodology and scientific explanation. Moreover, as Salmon (1990:32) remarks, "the mere recital of just any set of preceding occurrences may have no explanatory value whatever, the narrative must involve events that are causally relevant to the explanandum if it is to serve as an explanation." Such a relationship is too frequently a subject of assertion rather than demonstration.

The record of the hominid past is an unobservable to be discovered, recognized, and appropriately investigated in its own right. Although there have been significant, indeed remarkable, advances in documentation of the fossil and attendant archeological records, there clearly remains scant if any representation, and hence knowledge, of some potentially important and certainly relevant spatio-temporal realms. This is particularly reflected in deficits in documentation of (a) the earliest Pliocene of Africa (only recently has the recovery of *Ardipithecus ramidus* opened this vista); (b) the initial (immigrant) hominid inhabitants of Eurasia; and (c) some substantive gaps in the Pleistocene record in some major areas and particular time spans in Eurasia as well as Africa. As a consequence of fundamental primary data deficiencies, investigators are limited in respect to the kinds of issues that may be addressed, in the ability to effect comparisons, and of course in respect to the recognition of issues as yet wholly inevident ('unobservables').

Although the import of newly discovered documentation is often down-played, even dismissed, the ultimate impact of such acquisitions are in fact overall momentous, and not infrequently profound. They are often surprises and may well eventuate in that 'essential tension' that leads to entirely new and unforeseen perspectives. The lag between such instances and their consequent impact is often substantial, as has been documented repeatedly in the growth of paleoanthropology. It is also worthwhile to stress that, in part as a consequence, subject matter and problems undergo examination and investigation from different perspectives, varying methodologies, and unlike disciplinary frameworks such that the results are attended, oftentimes, by varying degrees of acceptance, skepticism, outright rejection, and associated controversy.

Many disciplines bear upon and potentially contribute ultimately to paleoanthropological endeavors. Each has its specialists and practitioners, and there has been an ever-accelerating trend toward multi- and cross-disciplinary research undertakings, particularly in field-focused efforts of exploration, data recovery, and consequent laboratory analysis. Despite such important and necessary progress, there frequently remains a notable lack of cross-disciplinary comparison, analysis, and integration requisite to the maximization of such particular research endeavors in regard to their bearing on fundamental issues in human evolutionary studies. Among others this is particularly manifest in respect to advances in chronometric dating, paleoenvironmental studies, paleogeographic reconstructions, taphonomic studies, analytical and distributional aspects of lithic assemblage studies, in the application of methods of cladistic analysis, and in the intensification of ontogenetic and biomechanical/functional studies relevant to the hominid fossil record. These disparities, coupled with differing reliances on and commitments to strongly contrastive methodologies and explicit (or implicit) frameworks of scientific explanation and hypothesis-formulation and testing are in no small part responsible for some major controversies and for attendant adversarial and polarized positions in respect to certain familiar problem areas across much of paleoanthropology.

Actually, developments in all aforementioned subject areas have experienced such significant advances as to impact importantly on the scope and nature of paleoanthropology. This is sufficient to transform it from the current emergent state into a full-blown, individuated discipline, having its own parameters, goals, procedures, methodologies, problem areas, and central issues. Paleoanthropology is close to a paradigm state without yet having achieved it. Unfortunately, that condition is insufficiently defined due in part to some significant consensual faults, and in part to the lack of adequate, explicit integration as a consequence of still often fragmented, unfocused aspects of some very relevant endeavors.

A few remarks are offered here in respect to some subjects of chapters in this volume. In each instance there is an important historical background, sometimes an extensive one, and the current states of researches in such instances reflect not only increments in data acquisition but also the impact of newer methodologies which together have led to enhanced understanding and, sometimes, substantially altered perspectives.

It has long been considered that the extant great (anthropomorphous) ape genera are more or less closely related and together form a higher taxon, either at the familial (Pongidae) or subfamilial (Ponginae) level. Similarly, the close affinity of the two African apes (*Pan, Gorilla*) vis-à-vis the distinctive orangutan (*Pongo*) has been widely accepted for well over a century (see Schultz 1936). This last conclusion has been extensively supported in the past thirty years by numerous biomolecular investigations, of several systems, that not only affirm this view but also reveal the rather unexpected close resemblance (affinity) of the African taxa to the human species (traditionally Hominidae) rather than the more distant Asian taxon. Consequently, the aforesaid familial/subfamilial rank must be disassembled, and the African ape-human lineages must be conjoined in a taxon of higher rank (Andrews 1992). Thus, and according to requirements of priority rules, family Hominidae includes extinct and extant apes and humans; African apes (Gorillini) and humans (Hominini) are joined in the subfamily Homininae; orangutans (Pongini) and extinct relatives (Sivapithecini) are joined in the subfamily Ponginae; other extinct ape groups also constitute tribes (3, at least) in another subfamily (Dryopithecinae). Substantial and repeated claims from some dozen molecular studies, thoughtfully discussed and evaluated recently by Bailey (1993), have been made for an immediate close affinity of chimpanzee (*Pan*) and human (*Homo sapiens*), relative to gorilla, which if validated would place both within Hominini. Others, including Marks in this volume, have been equally adamant in their insistence that the available evidence is still inconclusive and that the branching arrangement among these three taxa constitutes a still unresolved trichotomy. This happens to represent my own persistent view of such evidence, which is reinforced by a wealth of ontogenetic observations (Hartwig-Scherer 1993) reflecting the distinctive growth pattern of *Homo* relative to that shared by *Pan* and *Gorilla*. However, increasingly the closeness of *Pan* and *Homo* is being convincingly forced upon us as a consequence of intensified biomolecular studies. The extant African ape trio remains the essential outgroup for comparisons, cladistic or otherwise, for Hominini. However, the view that 'the chimp' is 'the model' for ancestral Hominini is still inferential and, until now, lacks adequate support from the hominid fossil record (which is almost non-existent between 8 and 4.5 Ma).

The denomination of *Australopithecus africanus* by Raymond Dart in 1925 was both unwelcome and severely criticized by a concerned majority of the scientific establishment. Broad, but non-unanimous, acceptance of such an extinct 'man-ape' or 'ape-man' into Hominidae (= Hominini) required over a quarter century to achieve. How can a situation such as this, constituting such a paradigm-enhancing contribution, be explained? In spite of some critical analysis of this event, it has still to receive in-depth study, and any explanation can only be attributed broadly to the (unspecified) realm of the existential. Nonetheless it can be argued that the validity of the genus was demonstrated within five or so years, and that its polyspecific nature was demonstrated by a decade later. Two further decades were required both to validate the proposal of multispecific taxa and simultaneously to reveal the sagacity of Robert Broom in his proposal (in 1938) of a separate higher taxon (*Paranthropus*) to accommodate the unparalleled range of variation in the then known hypodigm. In fact it would require another quarter century before the *Australopithecus/ Paranthropus* dichotomy would receive the concerted and serious consideration it merited for so long. Clarke offers here a welcome overview, from historical, morphological, and adaptational perspectives, of the australopithecine radiation, for indeed it was such, from his own deep familiarity with the primary fossil collections. It is now both misleading and inappropriate to consider, as was once commonly done, such hominid(s) as the root or link in the hominin lineage. There are minimally three well-defined lesser taxa (species) of *Australopithecus*, and upwards of five such of *Paranthropus*. In each instance lesser taxa have disjunctive distributions between the eastern and southern African subregions. Several distinctive taxa considered referable to *Homo* are now well documented as having co-occurred, evidently sympatrically, with *Paranthropus* species in both subregions.

For many decades evidence of very early representative(s) of *Homo* remained wholly unknown. In fact, claims for great, even Pliocene, antiquity of this genus have been repeatedly advanced and as often vigorously rejected by some then – influential members of the scientific establishment for upwards of a century. A growing understanding of late Cenozoic biostratigraphy and of the earlier Paleolithic some forty years ago indicated the absence of hominin documentation in Eurasia until well into the mid-Pleistocene; the record in Africa was dominated by australopithecines (in the south) and a few purportedly *H. erectus*-like fossils (in the Maghreb). The subsequent two decades witnessed the intensification of researches in the Olduvai basin and, subsequently, the initiation of protracted, large scale field efforts in the greater Turkana Basin. This led to the recovery, recognition, and denomination of *Homo habilis* in an isotopically dated context at Olduvai and its further confirmation in the Turkana basin, along with increasing numbers of at least one other taxon clearly representative of another, further derived representative of *Homo*. Together these discoveries have afforded a wealth of hominin documentation in the 2.0-1.4 Ma time span, forced re-examination of emerging conceptions of variability, taxonomy, and potential phylogenetic affinities of these newly acquired fossil samples, and initiated ever-expanding research programmes focused on the most ancient Paleolithic occurrences documented anywhere in the world. Wood discusses here this fossil record and evidence it affords for a multiplicity of ancient African *Homo* taxa (*H. rudolfensis, H. habilis*, and *H. ergaster*), for attendant extinctions, and implications for greatly enhanced insight into the roots of subsequent hominin evolutionary developments.

An extra-African dispersal of hominins into Eurasia became an important tenet of paleoanthropology within recent decades due to a much enhanced African fossil record of the better delineated Pliocene interval. The most approximate time for this extensive range expansion appears now to have been broadly coincident with or just after the Olduvai (N) subchron (sc), ~1.96-1.78 Ma. Two occurrences in western Asia, Erq el Ahmar (southern Jordan valley) and Dmanisi (Georgian Caucasus) seemingly constitute the nearly first manifestations of the dispersal event, which is very well-documented subsequently at 'Ubeidiya (central Jordan valley) at ~1.4 Ma. Recent isotopic (Ar/Ar) age assessments (Swisher *et al.* 1994) of fossiliferous (and hominin) occurrences in Java, at Perning (~1.8 Ma) and at Sangiran (~1.66 Ma),

coupled with redefined paleomagnetic evaluations, indicate an unexpectedly enhanced antiquity for hominin penetration into Sundaland, apparently toward the close
of the Olduvai sc, and certainly not long thereafter. In fact there is now purported
temporal overlap between those occurrences and the temporal range (in eastern
Africa) of both *H. ergaster* and *H. habilis*. The Perning hominin occurrence lacks
taxonomic substantiation, whereas at Sangiran, in spite of some strong opinions to
the contrary (Kramer 1993, 1994; Kramer & Koenigsberg 1994), multiple hominin
taxa appear very likely to be represented early on. Clearly such a dispersal 'event'
requires reconsideration and more explicit formulation, and full description and
in-depth comparative analyses of appropriate Javan hominin specimens from the
oldest reaches of that fossil record are sorely needed. Similarly, the seemingly belated
appearance of hominins in continental eastern Asia, perhaps some 1.0-1.2 Ma, now
requires fuller analysis and explication. Hominin occupation of this or still greater
antiquity has been claimed at times, and even now, for Europe. However, incontrovertible evidence to support such a perspective has not been readily forthcoming in
spite of the expenditure of much effort in the search. Personally, I consider that some
claims for such antiquity are overstated and as yet unconfirmed. However, others
may well be valid, both in southwestern and in central/eastern Europe, and I remain
more sanguine than some others (Roebroeks 1994; Roebroeks & van Kolfschoten
1994) about the potentiality, even probability, of hominin presence in the southern
reaches of Europe in the late Matuyama chron, broadly 1.5-1.0 Ma. This view is now
supported by the newly documented, and incontrovertible conjoint faunal/artifactual
occurrence in the Baza sub-basin of the Guadix-Baza intramontane (Betic system)
basinal complex of Andalusian Spain, almost comparable in antiquity to that of
'Ubeidiya in the Levant. Nevertheless, much fuller documentation is still required
for verification of some claims, and ultimately only new occurrences, in incontrovertible and dateable contexts, will enable resolution of this issue. It remains true,
certainly, that the most substantive and informative body of evidence falls within the
Brunhes (N) chron (*i.e.*, < 0.78 Ma).

Neontologists tend to view evolution from the top down (present to past) and
paleontologists from the bottom upwards (past to present). In each instance the
present (and fully 'known') is given and the past is largely 'unknown'. Neontologists,
because of their focus on 'known' and knowable, have quite different expectations
with respect to the unknown (and unknowable) past than the paleontologist, im-
mensely reliant upon and grateful for whatever is at hand at his/her particular moment
in time. The former can never be sufficiently satisfied by the evidence afforded by
the past, whereas the latter is continually reassured as such evidence as there might
have been comes to be enhanced. Although sharing a common goal, focused on
comprehension of evolutionary pattern and processes, their perspectives and approaches are recognizably unalike and their consequences are correspondingly different. Such distinctions are relevant in regard to human evolutionary studies, and in
particular to treatment of the documentation afforded by the hominin fossil record.
There is an accumulated historical baggage of phylogenetic inference and attendant
scenarial speculation that can only be considered now as of little value and scant

meaningfulness, in the same way as we might consider pre-Newtonian mechanics and pre-Keplerian planetary astronomy. T. H. Huxley long ago remarked that "the greatest tragedy in science is a beautiful hypothesis murdered by an ugly fact." And John Dewey put the matter this way:

> Old ideas give way slowly, for they are more than abstract forms and categories. They are habits, predispositions, deeply engrained attitudes of aversion and preference. Moreover, the conviction persists – though history shows it be a hallucination – that all the questions that the human mind has asked are questions that can be answered in terms of the alternatives that the questions themselves present. But, in fact intellectual progress usually occurs through sheer abandonment of questions together with both of the alternatives they assume — an abandonment that results from their decreasing vitality and a change of urgent interest. We do not solve them: we get over them. Old questions are solved by disappearing, evaporating, while new questions corresponding to the changed attitude of endeavor and preference take their place. (Dewey 1910:19)

One might thus be reminded of Planck's depressing dictum that a "new scientific truth does not triumph by convincing its opponents and making them see the light, but rather because its opponents die, and a new generation grows up that is familiar with it" (Planck 1950; see Hull 1989:ch. 3). In fact such is not necessarily or frequently the case, as any even cursory examination of much history of science reveals other factors, perspectives, and empirical resources frequently play a decisive role.

In recent years investigators have confronted anew the issue of species and clades in the hominin fossil record. The difficulty of employing the otherwise broadly accepted and widely employed biological species concept in paleobiology is commonly acknowledged. Some version of a phylogenetic (evolutionary) species concept is frequently proposed toward resolution of this issue. Such might include E. O. Wiley's (1978:18; 1981:25) reformulation of G. G. Simpson's evolutionary species concept as "a lineage of ancestral – descendant populations which maintains its identity from other such lineages and which has its own evolutionary tendencies and historical fate" (essentially a biological species concept, invoking reproductive continuity/individuality through time). The lineage is construed as "one or a series of demes that share a common history of descent not shared by other demes." The phylogenetic species concept (of Eldredge & Cracraft 1980:92; Cracraft 1983:170; Cracraft 1989:35) is considered "an irreducible (basal) cluster of organisms diagnostically distinct from other such clusters, and within which there is a parental pattern of ancestry and descent." Operationally, this implies diagnostic, heritable character(s) and reproductive cohesion within clusters (demes or related, larger aggregates).

The critical component in most such delineations is the deme, considered "as the unit of natural history and of evolutionary divergence," and defined as "a communal interbreeding population within a species" (Carter 1951:142), or, following Mayr & Ashlock (1991:413), "a local population of a species; the community of potentially interbreeding individuals at a given locality." The term was introduced by Gilmour & Gregor (1939) for a local interbreeding population or community ("any assemblage of taxonomically closely related individuals"), and distinguishable by reproductive (genetic), geographic, and ecological (habitat) parameters. Together, demes (sometimes as isolates) constitute subspecies, the "aggregate of local populations of a species inhabiting a geographic subdivision of the range of the species. . ." (Mayr & Ashlock 1991). It is well known that much of the hominin fossil record is characterized by fragmentary and otherwise incomplete skeletal parts such that even an individual, drawn of course from a deme, may be quite imperfectly known. Important (and notable) exceptions are those singular collections of multiple and more extensively preserved individuals – as instanced among *Australopithecus* species (Hadar locality A.L. 333, Sterkfontein-Mb. 4), *Paranthropus* species (Swartkrans-Mb. 1), *Homo erectus* (ZKD-1, Ngandong), Neandertals (Krapina) and antecedents (Sima de los Huesos, Atapuerca), and *Homo sapiens palestinus* (Skhul, Qafzeh) – that afford significant demic samples. Such are of course spatio-temporally disparate, regionally localized samples of paleodemes (p-demes). Due to the disjunctive distribution of late Cenozoic continental sediments, in whatever geological setting, it is rare indeed to have much spatial (geographic) documentation of subspecies distribution and variation from a suitable series of p-demic samples.

P-demic samples are the basic stuff of the hominin fossil record. Trinkaus (1990) has also considered that "the best approach . . . is probably one that regards the available fossil samples (or specimens) as representatives of prehistoric populations or lineages acting as portions of dynamic evolutionary units." A central problem has been, and remains, that of bridging the gap between individual specimens and such p-demic samples to achieve an appreciation of the subspecies and its variation. Finds from the earlier history of human paleontology, including types and/or less complete, fragmentary specimens always require re-evaluation in the light of fuller information afforded by both more complete and numerous remains recovered subsequently. This obvious choice is not always sufficiently appreciated or adequately pursued (however common practice in paleontology). After an early history and long persistence of species recognition and taxonomic splitting (Campbell 1965), human paleontology has often been dominated by an overriding concern with subspecies and attendant reluctance to confront the issue of species lineages. Thus, there is the further problem of the delineation of species lineages on the basis of evidence afforded by p-demic samples, however substantial or (more often) limited. This is a problem faced by paleontologists for generations, but one too often evaded by students of the hominin fossil record.

Ian Tattersall (as summarized in this volume) has been most vocal in his opposition to such practice. He has observed, rightly in my view, that "In any group other than Hominidae the presence of several clearly recognizable morphs in the record of the middle to upper Pleistocene would suggest (indeed, demonstrate) the involvement of several species" (Tattersall 1986:170). Thus, to him, "The interpretation of most human fossils subsequent to about 0.5 myr as belonging somehow to *Homo sapiens* is perhaps the most comprehensive smoke screen of all, and could only have been maintained by a zealous refusal to consider characters other than brain size" (Tattersall 1986:169). And, he further concludes "that there is neither theoretical nor practical justification for continuing to cram virtually all fossil humans of the last half-million years into the species *Homo sapiens*" (Tattersall 1992:348). I would add that the whole of the history of genus *Homo* is increasingly being viewed by some (thankfully few) students of human evolution as remarkably free of the usual events — adaptive radiation, diversification (including cladogenesis), divergence, stasis, endemism, extinction, dispersal and range extension, etc. — upon which paleobiologists commonly focus their attention. Certainly to many of us, this suggests that some perspectives and certain hypotheses concerning the evolution of genus *Homo* are far wide of the mark. There is scant reason to marvel over the lack of concurrence among investigators so long as such profound and fundamental divergences continue to be manifest.

Some frameworks of hominin evolution strongly smack, if not of orthogeneticism, at least of outright progressivism. Gould (1988:319) considered evolutionary progress "a noxious, culturally embedded, unstable, non-operational, intractable idea that must be replaced if we wish to understand the patterns of history." If not overtly, at least covertly the usage of stage and grade concepts persist, the latter leading surely at times to paraphyletic groupings. Familiar ascriptive terms in human paleontology, aside from 'primitive,' include 'archaic' (adj., characteristic of a much earlier, often primitive period); 'intermediate' (adj., lying or occurring between two extremes, in a middle position; a gradistic term, frequently employed as a noun and conflated with intergradation); 'transitional' (adj., passage from one form, state [stage] to another, and similarly gradistic); and, faux de mieux, 'anatomically-modern' (as the redundant compound noun, 'anatomically-modern human'). All such terms would seem to have no effective role in the vocabulary of (human) evolutionary biology. Where is the study of historical, evolutionary events? Of branching sequences? Of ancestral morphotypes and their derivatives? Of structural complexes, their functional analysis, and their transformations? Of genetic isolates and p-demes? Is there no role for endemism? Indeed, are current studies of human evolutionary biology informing us realistically of the pattern of hominin phylogeny? Alas, the only response is too often in the negative.

Set out in the remainder of this chapter are several issues that particularly merit re-examination from new or at least different perspectives if elucidation of hominin phylogeny is a central goal of human evolutionary studies.

Primary Dispersal(s) of *Homo* Species

The nature of initial, ancient Eurasian hominins and their presumed African source are critical components in elucidating the subsequent phylogeny of genus *Homo*. The African source is now generally acknowledged to have been *H. ergaster* (= *H.* aff. *erectus* of some authors), the known age range of which, in East African rift localities, is 1.85-1.5 Ma. The oldest known extra-African hominin occurrence is in western Asia, at Dmanisi (Georgian Caucasus), in a purportedly comparably ancient situation (Olduvai [N] sc) (Gabunia & Vekua 1995). While admittedly sharing some (presumably) primitive features with African (Turkana Basin) analogues, Dmanisi is demonstrably derived, both gnathically and dentally, and is thus unique.

A hominin locality in eastern Java, that of Perning, is now considered of broadly similar antiquity (1.81 Ma), and another central Javan locality (at Sangiran dome), with an age of 1.66 Ma, falls similarly within the temporal span of the African species (Swisher *et al*. 1994). The distance between the African and Indonesian Sunda Shelf occurrences is 76° of longitude. At Sangiran some hominin cranial and gnathic remains not originally attributed to *H. erectus erectus*, demonstrably provenienced to the Grenzbank (conglomerate) and underlying Pucangan clays, apparently diverge markedly and fundamentally from penecontemporaneous African *H. ergaster*. However, some material from the same geological horizon(s) is commonly attributed to *H. erectus*. If, as seems most probable, multiple hominin taxa are represented in the eastern (Sundaland) extent of this range, at ~1.6 Ma, is it appropriate to view *H. ergaster* as the single source of such taxa and, as well, of all subsequent hominins? To what extent should effects of insularity and faunal filters be considered in respect to the initial peopling by hominins of the Sunda Shelf, as well to their subsequent evolutionary history?

There is no such documentation in the whole of Europe, and still scant documentation even of hominin presence there prior to ~1.0 Ma. Were perhaps subsequent African populations the source of European hominins, particularly as (Asian) *H. erectus* has, as yet, not been documented to occur in Europe? Some phenetic resemblances have at times been noted between some hominin samples from the earlier mid-Pleistocene time range of southern Europe and others from sub-Saharan Africa. Are such purported resemblances real, and to what extent is this indicative of genetic affinities as opposed to homoplasy?

Students of hominin evolution have been chary of the consideration of population dispersals, range extensions, constrictions, ebbs and flows, etc. Tchernov (1992) has recently discussed, albeit only briefly, aspects of dispersal and varied terminological usages in evolutionary ecology and biogeography. It would seem that paleoanthropologists have envisioned a sort of one-time, one-way out of Africa scenario that set the stage for all subsequent (and attendant) hominin evolutionary events, with the dispersal, perhaps really a sort of persistent, accelerating range extension, presumptively expanding so as to fill all appropriate habitats, overcoming or by-passing formidable barriers, rather like filling an empty saucer. However, might not the process have been more akin to filling a muffin pan or egg carton, being comprised of more limited and restricted or compartmentalized intrusions leading to patchy, even sporadic and impersistent, distributions, differentiated by barriers in space and through time?

Demes — Continuities, Discontinuities, and Pleistocene Hominin Clades

Within species patterns of variation reflect the summation of subspecific variabilities as measured across demic samples. The biological species concept is not immediately applicable to fossil samples; demic and subspecific entities are of course represented in the fossil record, as in extant species. Thus, as Shea *et al*. (1993:281) have strictured, "The most rigorous course open to the paleobiologist is to indirectly test patterns and distributions within the fossil sample against those derived from careful analysis of *both species and subspecies* of close phylogenetic relatives and groups occupying similar ecological and adaptive zones" [italics in original].

As Harrison (1993:362) points out, it is "important to identify species groupings as an initial step in a cladistic analysis, and not to use cladistics as a method to identify species groupings." Thus, because "species can only be identified in the fossil record in terms of morphological criteria, and although there are no absolute rules for the recognition of species boundaries, intuitive ideas of the range of well-known extant groups provide models for the inclusion or exclusion of individuals." Hence, "fossil species are recognized by employing entirely phenetic concepts based on analogs derived subjectively from morphological ranges of variation seen in modern species." Nonetheless, and as frequently acknowledged, strict reliance on degree of (phenetic) resemblance may be hazardous and even misleading. Simpson (1963) thus noted that, "As a rule with important exceptions, degrees of resemblance tend to be correlated with degree of evolutionary affinity. Resemblance provides important, but *not the only*, evidence of affinity" [italics in original].

Multivariate studies of craniometric variation in extant large-bodied anthropomorphous apes "discriminate in statistically significant fashion among the three primary groups (or multiple local populations) of *P. troglodytes*" (Shea *et al.* 1993), with Mahalanobis D^2 values ranging between 1.416-2.166 (or 1.79-3.81 in a prior analysis), whereas in *G. gorilla* D^2 values between demes of three different subspecies are 2.01-4.46 (mean, 3.51) and between demes of the same subspecies are 1.37-2.29 (mean, 1.76) (Albrecht & Miller 1993). Howells' (1989; also 1973) extensive multivariate analyses of craniometrics in 28 modern human samples afford invaluable in-depth documentation of within-group and between-group variability and enable various levels of cluster analysis and distance assessment between samples, geographic areas, and additionally, numbers of late prehistoric and subrecent samples. Howells' work demonstrates that recent humans are "cranially rather homogeneous, with a limited dispersion of character" (1989:16); he states, "the impression, true or not, is that the modern unity is fairly recent" (1989:70). Examples of regional (including intracontinental) D^2 values are those among three African samples, 4.3-5.4, and between such and Bushmen, 5.7-6.8; among three Australasian samples, 4.1-5.1, and between such and Andamanese, 6.6-7.5; and among four East Asiatic samples, 2.9-4.0 (2.6 between two Japanese samples), and between such and Ainu, 4.9-5.4. Minimum D^2 distances recorded were within Europe, 2.9; maximum distance recorded was 9.9 (between north Asian, Buriat, and Australian aborigines). The latter is similarly reflected in the dentition differences recognized between Sinodont/Sundadont and Australasian population aggregates.

Simpson (1963; also 1961) made important distinctions in respect to the language employed in taxonomy and different designative words (names, terms) and their kinds, with particular regard to processes, name sets, and referents (see Figure 1). He especially noted, in reference to hominin evolutionary studies, that "one of the greatest linguistic needs in this field is for clear, uniform, and distinct sets of N_1 and N_2 designations, applied to specimens and to local populations as distinct from taxa." It is well known that innumerable workers concerned with hominin paleontology

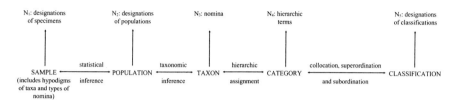

FIGURE 1. "Schema of processes (arrows), name sets (N), and references (capitals) in taxonomy. Vertical arrows all represent the processes of designation or symbolization. The processes represented by horizontal arrows proceed logically from left to right, but in practice no one operation can be carried out without reference to the others. These arrows are therefore drawn pointing both ways." (Simpson, 1963.) (Permission to reproduce courtesy of the Wenner-Gren Foundation for Anthropological Research.)

have been notoriously imprecise in their usages, often skipping confusingly between what are in fact distinct entities and concepts. The specification of different name sets (N) is requisite and mandatory for any measure of consistency and progress in this endeavor. Following Simpson the essential distinctions are: N_1 = specimen (object) designation (locality name, repository/catalogue numbers; = hypodigm-1); N_2 = group designations, population (paleocommunity or p-deme in author's sense) designation, reflecting inclusive (genealogically) related geographic group(s) (= hypodigm-2); N_3 = formal nomen or taxon, composed of N_2 hypodigms as drawn from and representative of subspecific and specific taxa (= hypodigm-3); N_4 = ranking designation, categorization within a hierarchic (Linnaean) classification. It should be stressed of course that the single (N_3) type, referring to a chosen holotype or, failing that, a subsequently designated lectotype, functions as a name bearer (or onomatophore, in Simpson's one-time usage) for a particular biological entity. It is not, and cannot function as, reflective of a full or inclusive spectrum of character states or their variations, and so on. However, N_2 designations are indeed revelatory in those respects. I consider that the N_2 designation is greatly undervalued and under-employed, yet as Simpson states "it is often necessary to recognize and designate a local population that is a part of a taxon but does not in itself comprise a whole taxon," and, thus, "for some populations a different set of names or symbols, N_2, may therefore be required." In the fossil record such p-demes exhibit bounded-ness, spatio-temporally.

Upwards of twenty such p-demes are distinguishable within the immediate ancestry subsumed within *Homo*, exclusive of the *H. habilis* and *H. rudolfensis* taxa, and exclusive of, in the strict sense, *Homo sapiens* (= 'anatomically-modern' humans of authors). It is to such p-demes, or some members thereof, that much comparative and analytical study has been devoted, although the demic focus has not always, even often, been made sufficiently explicit.

The Principal Distinguishable African Hominin P-demes Are:

Nariokotome. Fortunately there are three partial skeletons within this group: a subadult skeleton (WT-15000); another attributed (pathological) partial skeleton (ER-1808); and the first discovered example, comprising associated cranial parts, mandible, and some postcrania (ER-730). Crania (ER-3733, -3883) and adult (ER-992, the type of the nomen, *ergaster* Groves & Mazák) and subadult (ER-820) mandibles are well-represented, and there are as well a number of referred and varied postcrania. A representative of this group is also documented at Konso-Gardula (southern Ethiopia). This deme is probably also represented (partial cranium SK. 847, etc.) at Swartkrans (Mb. 1 units). *Paranthropus boisei* (in eastern Africa) and *P. crassidens* (in southern Africa) are sympatric australopithecine taxa at all known localities/temporal intervals.

Olduvai/LLK-II. The calvaria of Olduvai Hominid 9, of rather uncertain stratigraphic provenance, is the principal specimen. It is the type of the nomen, *leakeyi* Heberer. A smaller, partial cranium (VEK-IVa-12) exhibits comparable morphology. Direct associations of mandibles and crania are still unknown, but two partial specimens (hemi-mandible O.H.-22, and fragment O.H.-51) are possible candidates for this p-deme. The only likely postcrania (innominate, femoral diaphysis) may be those of WK-IV(a), O.H.-28.

Tighenif (Ternifine). This p-deme is, for now, recognizable only in the northwest African Maghreb. The cranium is unknown except for a parietal, whereas three mandibles with dentition are well represented, as are some nine isolated teeth, all from the eponymous locality. A mandible (T-1) is the type of the nomen, *mauritanicus* Arambourg. In known elements there are both (derived) distinctions from, and some resemblances to, the Olduvai/LLK p-deme hypodigm — but comparisons are weakened by either insecure attributions (at Olduvai) and/or poor cranial representation (Tighenif). A single postcranial (femoral diaphysis) might be considered, parsimoniously, as representative of this p-deme (from Ain Maarouf = El Hajeb). Probably attributable to a similar, and thus presumably related, p-deme are geologically younger gnathic-dental remains from Sidi Abderrahman (Littorina Cave-F; Thomas Quarry-1) and calvarial (frontal) and dental remains (Thomas Quarry-3). The Salé partial cranium unfortunately affords a most incomplete and imperfect characterization of calvarial morphology due to its quite extensive (occipital) pathology; its demic affinities are obscure.

Kabwe. The type of the nomen *rhodesiensis* Woodward is represented by a complete adult cranium (Kabwe – NNM. E-686). It is still the most complete individual cranium of this p-deme. Other individuals are represented by partial maxilla (K-2, E687) and parietal (K-3, E897); some three individuals are represented by a lot of fragmented parts of the axial (sacrum, innominate) and upper (humerus) and lower (femora, tibia) appendicular skeleton, of uncertain association and provenance. An additional calvaria from Elandsfontein corroborates closely the morphology of the first found individual. Additional representatives probably include the Ndutu cranium, and only conceivably Olduvai Hominid 23 (Masek beds), a mandible fragment.

Tentatively attributed to this group is the Bodo (Middle Awash) cranial specimen, which shares most derived features with Kabwe and but few with the Olduvai/LLK-II individual. The extremely fragmentary condition of Eyasi-1 leaves uncertain its potential affinities in respect to this group. Hominin remains from Melka-Kontouré (Ethiopia) and from Baringo-Kapthurin Fm. (Kenya) might fall into this group or ultimately prove to constitute a separate p-deme.

Irhoud. This p-deme is well-typified by adult cranium (no. 1) and subadult calvaria (no. 2) and mandible (no. 3); there is also an imperfect infant humerus (no. 4), all from the eponymous cave filling. Partial cranial and gnathic-dental remains from Kébibat (Mifsud-Giudice quarry) may well represent the same or closely-related p-deme. A fronto-facial fragment from Zuttiyeh (Galilee) is attributable to the same demic group.

Florisbad. Although less complete than some other specimens, this partial cranium is demonstrably distinctive, and in fact constitutes the type of the nomen, *helmei* Dreyer. Other cranial specimens now known to be closely similar and hence attributable to this demic unit include Omo/Kibish-2 calvaria, Eliye Springs (ES-11693) partial cranium, and Laetoli (Ngaloba) H.18 cranium, the latter the most fully-preserved of the lot.

KRM. This p-deme is based on fragmentary but varied cranio-gnathic-dental elements and postcranial parts from several levels of SAS (Sands-Ash-Shell) Member (up to 10 m, with MSA II industry) of Klasies River Mouth Main cave complex. The complex is a single sedimentary depository subdivided spatially into Caves 1, 1C, 2, 1A, and adjacent 1B. Most remains derive from the lower such levels of Cave 1, with fewer specimens from 1A (upper levels) and from 1B (basal levels); there is some evidence to suggest most such specimens may well be of broadly comparable antiquity. The MNI is at least 5. There is some evidence to suggest that fragmentary maxillary parts (two individuals) from the lower and older LBS (Light Brown Sand) Member of 1A represent an antecedent but related population. It can be argued, both pro and con, that the hominin sample from Border Cave should best be attributed to the same, or at least a closely-related, p-deme. I tend to support that view. Surprisingly, requisite in-depth comparative morphometric studies appropriate to adequately test this hypothesis have yet to be carried out. The same might be said in respect to the (largely dental) samples from Die Kelders and Equus caves, and those vault and teeth parts associated with MSA III industry at KRM-1A (South Africa) and from Mumba shelter (Eyasi basin, Tanzania), respectively.

The Principal West Eurasian Hominin P-demes Are:

Dmanisi. The recent discovery in the Georgian Caucasus of a well-preserved hominin mandible, with dentition, overlying the Mashavera basalt, with important faunal and artifactual associations is a major contribution toward elucidation of the peopling of Eurasia. The mosaic morphology of the single specimen, including its derived characters in comparison with African counterparts, indicates its unique p-deme status.

Atapuerca-Gran Dolina. Very recent excavations in the lower infill of the cavern Gran Dolina of the Atapuerca (Spain) karstic complex has afforded unsuspectedly ancient hominin occupation, just prior to the Brunhes/Matuyama paleomagnetic reversal event. The occurrence in level TD-6, which includes an artifactual assemblage and faunal residues, comprises thus far cranial, gnathic, dental, and some postcranial parts of some 5-6 hominin individuals, subadult and adult. Like the associated mammal remains, there is evidence of breakage, impact fractures, and incision marks, all suggestive of anthropophagy. The morphology, as thus far revealed, is distinctive and seemingly quite divergent from less ancient European counterparts. The sample, surely to be augmented in future excavations, merits recognition as a new p-deme. An incomplete hominin calvaria from Ceprano (Sacco-Lire basin, Latium), from a situation low in the known local succession, may well prove on fuller study to constitute another representative of the group.

Mauer/Arago. Hominin remains from these localities constitute the next oldest documented samples in Europe proper, to which now Boxgrove, Sussex (a tibial diaphysis) may probably be added. The isolated Mauer mandible with dentition is the type of the nomen, *heidelbergensis* Schoetensack. The Arago sample, comprising cranio-facial elements, two partial mandibles, dentitions, and postcranial elements, is best taken as basis of comparison. Overall this p-deme is distinguished by an idiosyncratic (regional) mosaic of some (sym)plesiomorphic cranial, gnathic, and postcranial features coupled with derived (apomorphic) features of other aspects of cranium (fronto-parietal elements, facial skeleton) and mandible (symphysis and ramus, dentition) which may approximate structures characteristic of subsequent Neandertals. The extent to which fragmentary hominin remains from Italy — including Fontana-Ranuccio (dentition), Visogliano-2 (mandible fragment), and Venosa-Notarchirico (femoral diaphysis) — might reflect this or another related p-deme remains to be evaluated.

Petralona/Atapuerca (Sima). This p-deme is among the best known and characterized in the European mid-Pleistocene, initially as a consequence of the recognition of the distinctive, beautifully preserved Petralona (Khalkidhiki) cranium, the type of the nomen, *petralonensis* Murrill. More recently, the extensive hominin assemblage of almost all major skeletal elements from several dozens of individuals recovered from Sima de los Huesos, Atapuerca, has afforded the largest single hominin sample from the mid-Pleistocene of Europe. This affords also a basis for fuller evaluation and attribution of isolated or incomplete/damaged specimens (Montmaurin-La Niche mandible, Vértesszöllös occipital and dental elements, and Steinheim, Swanscombe, and Bilzingsleben crania) about which considerable discussion and indeed controversy prevailed over too many years. The Steinheim specimen is the type of the nomen, *steinheimensis* Berckhemer, and this of course has priority over *petralonensis*. These examples, along with the still ill-known Apidima Diros (Greece) crania and, perhaps, the recently found and remarkably complete Altamura (Italy) skeleton, may ultimately prove to be attributable to the same demic group. Similarly, a diversity of more or less fragmentary specimens from Italy — in Liguria, Prince Cave (innominate), in the Latium, Castel di Guido (cranial, femoral fragments), Cava Pompi (postcrania, cranial fragments), Casal de' Pazzi (cranial fragment), Sedia del Diavolo (postcranial/cranial fragments), and Ponte Mammolo (postcranial fragment) — require comparative examination from this perspective. This p-deme reveals persistence of some plesiomorphic features, but substantially stronger and widespread expression of Neandertal synapomorphies.

Neandertal. This p-deme was the first pre-modern human form known in the fossil record and is now, overall, the most fully and geographically extensively represented hominin from western Eurasia. The first recognized example, a partial skeleton from Feldhofer cave, Neandertal, is the type of the nomen *neanderthalensis* King. It is most extensively documented in the Last Glacial (^{18}O Stages 4-3) from the Siberian Altai, central (Teshik-Tash) and western Asia (Shanidar) into the Levant (Kebara, Amud, Tabun, Dederiyeh), the Crimea, west Caucasus, and throughout the reaches of middle and western Europe into the Apennine and Iberian peninsulas. Its earlier history is documented in Stage 5 (Krapina, Saccopastore, La Chaise – Bourgeois – Delaunay, Gánovce, Salá, Taubach, Salzgitter-Lebenstedt), Stage 6 (in France, Lazaret, La Chaise – Suard, Fontéchevade, Biache, and in Wales, Pontnewydd), and even into Stage 7 (Ehringsdorf). Partial or largely complete skeletons, some in definite interment circumstances, enable a very full, if not absolutely exhaustive, elucidation of skeletal paleobiology. The distinctiveness of the p-deme, with extensive autapomorphic character states, throughout its spatio-temporal range, and including variability, aspects of dimorphism, and ontogenetic development, is well-established and extensively documented. There are overall phenetic resemblances, and sharing of particular discrete traits, with antecedent demes, particularly with the Petralona/Atapuerca p-deme, but substantially less so with that of Mauer/Arago.

Skhul/Qafzeh. This p-deme is well-known, but only surely so from the eponymous Levantine cave localities of M. es-Skhul (Mt. Carmel) and Djebel Qafzeh (Galilee). These samples comprise partial or semi-complete skeletons, with associated skull parts, of 10 and 13 individuals respectively, including both adults and infants. (An infant individual, No. 1, from the es-Skhul site is designated lectotype for the species nomen *palestinus* McCown and Keith.) In cranial vault and facial morphology, gnathic features, some dental characteristics, both ontogenetically and in adults, and in varied aspects of postcranial (both axial and appendicular) morphology and proportions, these samples differ consistently and substantively from Neandertals and more closely approximate the *H. sapiens* condition, without however replicating it. Some phenetic, and thus presumably genetic, affiliations with the Maghrebian p-deme of Irhoud, of which a representative (Zuttiyeh) is known in the southern Levant, have been established. Potentially attributable to the Skhul/Qafzeh deme is Tabun-C2 (adult mandible), which like the eponymous localities is associated with a 'C-type' Mousterian of Levallois facies industry.

The Principal East Asian Hominin P-demes Are:

Yunxian. This stands, for the moment, as among the oldest documented hominins in eastern Asia. It is represented by two relatively complete but substantially distorted crania from Quyuanhekou, richly fossiliferous and artifact-bearing middle terrace deposits, Han River (Hubei). So far as known this p-deme shares many (not all) of the features characteristic of the Zhoukoudian p-deme, whereas others are attenuated if expressed at all. The Gongwangling (Lantian, Shaanxi) partial calvaria, including maxillary fragments, which is deformed with substantially altered surface relief, shares features both with this deme and that of Zhoukoudian, in the former instance thus apparently invoking a more plesimorphous state.

Zhoukoudian. This was the first major documentation of an extinct form of humanity in continental East Asia. The essential sample, to which the nomen, *pekinensis* Black & Zdansky applied, is the substantial hominin collection of Zhoukoudian-1 locality (Hebei), comprising some 45 individuals, 20% of which are immature, largely represented by calvarial, gnathic, and dental elements, and only few (14) and incomplete postcranial parts. The particular morphological distinctiveness of this deme is thus, overall, well-documented, and as a consequence it is a central comparative focus for all subsequent additions to the Asian hominin record. Thus, reputable records of the deme exist in other areas of largely northern/central China in Shaanxi (◆ Chenjiawo, a mandible, holotype of the nomen, *lantianensis* Woo), Shandong (˟ Qizian Hill), Henan (Xinghuashan), Hubei (Longushan, Builongdong, Longgudong), Anhui (* Longtandong), Kangsu (˟ Huludong) and Guizhou (Yanhuidong). Many such samples compromise isolated or associated teeth, whereas two (˟) comprise cranial parts, another (*) an association of skull and postcranial parts of some four individuals, and another (◆) an isolated mandible (and, perhaps, the oldest documentation so far of this p-deme).

Dali. This p-deme is represented, perhaps solely, by a well-preserved cranium with displaced facial skeleton from ancient terrace gravels of the locality Tianshuigou (Shaanxi), the name deriving from the eponymous county. This is the type of the nomen *daliensis* X. Wu, proposed to accommodate this and some other purportedly related specimens. This individual lacks some (not all) plesiomorphies and autapomorphies of both Yunxian and Zhoukoudian p-demes and, contrariwise, exhibits derived features of both cranial vault elements, their proportions and conformations, and, especially, distinctive reorganization of facial structure. Other localities which might, conceivably, afford representatives of this deme could include Walongdong and Guojujan (Hubei) and Chaoxian (Anhui), and perhaps also Dingcun (Shanxi); however, comparative studies are still required to substantiate such an inference.

Jinniushan. This p-deme, apparently distinct from that of Dali, is represented by a largely complete cranium (and dentition) and some associated axial and appendicular skeletal elements from the eponymous cave locality (Liaoning). Although as yet only preliminarily described, this individual departs overall in significant aspects of cranial vault and facial morphology from Dali, and thus exhibits a more derived condition. It is unclear, as yet, the extent to which this structure is paralleled or reflected as well in the Xujiayao skeletal series of cranial vault bones and teeth (sampling some 10 individuals) from the Datong basin (Shanxi).

Maba/Hathnora. This p-deme was first documented in 1958 by the recovery of a large hominin partial calvaria from a fissure fill in Shizi Hill (near Maba, Guangdong). It is type of the nomen, *mapaensis* Kurth. Its divergence from the Zhoukoudian morphotype, and otherwise purportedly archaic features, coupled with incomplete condition, long obfuscated elucidation of its affinities. The recovery of another partial calvaria near Hathnora, middle reaches of Narmada valley (Madhya Pradesh) from a terrace fill conglomerate affirms the morphological features of Maba. (This was the type of the nomen, *narmadensis* Sonakia.) Although unfortunately incomplete in critical areas of the vault, and lacking facial skeletons, both are fundamentally similar in total morphological pattern and exhibit a melange of primitive, derived, and unique features. These diverge from those of Zhoukoudian, Dali, or Jinniushan p-demes, and strongly suggest separate demic attribution.

The Principal Sunda Shelf Hominin P-demes Are:

Glagahomba/Sangiran. This designation is proposed for a poorly known (and ill-appreciated) hominin considered here as a distinctive p-deme. The name refers to a locality on the northern limb of the Sangiran dome (anticline) near which one such hominin specimen (Sangiran-8 or B, imperfect mandible with dentition) is specifically provenienced. The initial ('type') specimen (Sangiran-6 or A, right mandible fragment with some dentition; the lectotype of the nomen, *palaeojavanicus* von Koenigswald) derived, ambiguously, from this same general area, and more particularly to the north toward Brangkal. Another such fragment (Sangiran-5, reportedly collected east of Kalijoso, and the holotype of the nomen *dubius* von Koenigswald), of unspecified provenance, might also be attributed to this group, as has been one (Sangiran-31) and/or another (Sangiran-27, cranial base and upper dentition) calvarial/cranial remains, the former from the central dome area. These several examples have such a distinctive set of gnathic, dental, and vault morphologies and sizes, strongly autapomorphic, as to set them apart from other peninsular p-demes, and particularly that of Trinil/Sangiran.

Brangkal/Sangiran. This designation is proposed for a distinct p-deme, recovered in the northern area of the Sangiran dome, as represented by the partial cranial remains with maxillary dentition of Sangiran-4. This is the holotype of the nomen *robustus* Weidenreich. Its salient characters and known total morphological pattern were adumbrated fifty years ago. It is unclear whether isolated mandible portions with dentition — Sangiran-1b (and maxilla fragment 1a) from south Bukuran (eastern dome) and Sangiran-9 from near Tegopati/Wonolelo (northeastern dome) — should be similarly attributed or not. This is also unfortunately the case in respect to the Perning-1 (eastern Java) infant calvaria, type of the nomen *modjokertensis* von Koenigswald, and (perhaps) the oldest documented hominin on the Sunda Shelf.

Trinil/Sangiran. This p-deme is among the best represented of Pleistocene Javan Sundaland hominins. The holotype, a calotte (Trinil 2), excavated from Trinil (Solo River, east-central Java) is the source of the nomen *erectus* Dubois. Postcrania (femora), also excavated, have been both deleted from (no. T-3) and associated with (nos. T-6-9) this sample (which also includes a P_1). However, the first specimen found was an indeterminate mandible fragment from Kedung-Brubus, from which a femoral diaphysis was subsequently recognized. Anticlinally deformed lahars, volcaniclastics, and fluviatile sediments of Sangiran dome afford the largest single demic sample from this region — cranial elements (14, including two crania, two natural endocasts, nine gnathic elements with dentition), isolated teeth, and uncommon postcrania (1+), in an incomplete listing. The sample is distinguished by its extensive hyperostosis and suite of autapomorphic character states.

Ngandong. This p-deme is represented by the substantial excavated hominin sample from the eponymous locality of the central Solo river. It comprises five cranial vault elements and nine partial or more complete calvaria, adult and subadult, and several postcrania (tibial diaphyses, innominate fragment). This constitutes the type of the nomen *soloensis* Oppenoorth (of which a calotte, Ngandong-1, is the holotype). Two other localities of this drainage, Sambungmachan (a meander between Trinil and Sangiran) and Ngawi (east of Trinil), have each yielded a calvaria (and the former also a tibial diaphysis) attributable to this deme. It shares the same fundamental morphotype of the Trinil/Sangiran p-deme, associated with some derived states of calvarial size, proportions, and shape.

Of these twenty or so distinguishable p-demes only three have not received categorical ranking at some time and thus do not bear specific or subspecific nomina. Such recognition reflects, usually, the predominantly typological approach traditionally prevalent in human paleontology and the mind set to which Simpson directed attention. It led Ernst Mayr (1950) to recognize that two circumstances have attended such nomenclatorial plurality: "a very intense occupation with only a very small fraction of the animal kingdom which has resulted in the development of standards that differ greatly from those applied in other fields of zoology . . ." and "the attempt to express every difference of morphology, even the slightest one, by a different name and to do this with the limited number of taxonomic categories that are available." Mayr (1950) then suggested several 'practical rules': "not to assign a formal name to any local population or race that does not deserve subspecific rank"; "give trinomials to all forms that do not deserve higher than subspecies rank"; and "to group together as subspecies groups all those subspecies within a species that form either geographical or chronological groups." After such practical admonishments, he then unfortunately suggested that humankind "speciated only once if our assumption is correct that never more than one species of man existed on earth at any one time," thus laying the theoretical groundwork for the so-called 'single-species hypothesis' which ultimately confounded human evolutionary studies for nearly fifteen years. Since the corollary to this perspective was that, in this view, "the known diversity of fossil man can be interpreted as being the result of geographic variation within a single species of *Homo*," the stage was similarly set for the promulgation of the so-called 'Multiregional Evolution' model (or hypothesis), the roots of which extend back, via Dobzhansky (1944), some fifty years at least. Perhaps the single strongest longtime supporter (as successor) of this perspective was the late J. N. Spuhler (1993:291; also 1985, 1988, 1989) who considered that "the combined morphological, archaeological and molecular information explains the origin of modern humans by regional evolutionary transformation with continual unification of species by gene flow and local isolation by distance."

Over the past several decades multivariate statistical methods have been increasingly adopted in biological anthropology, and have come to play as well a prominent role in both analytical and comparative studies of the hominin fossil record. Thus, almost all of the aforesaid hominin p-demes, or at least some of their constituents, have been variously subjected to such treatment, often in conjunction with other such analyses of diversity and degrees of phenotypic similarity and divergence reflected among modern human populations. Table 1 (see Appendix A) is a non-exhaustive inventory of such studies, their foci, and the principal procedures employed.

The significant outcome of the totality of such studies is, according to the appropriateness of the procedure(s) and scope of morphometric parameters employed, that the distinctive character of particular samples has been well-delineated and verified, and the phenetic (and thus, inferentially, phylogenetic) affinities reasonably quantitatively elucidated. Moreover, the diversity exhibited by spatio-temporal samples of known hominin p-demes is unparalleled among samples of modern human populations, in which diversity is low whether assessed by morphological variability or conventional genetic markers. Thus, "the most mitochondrially different humans known are less different even than the only two siamangs sequenced to date or than lowland gorillas living in a restricted area of West Africa," and such "limited genetic diversity becomes equivalent via application of the molecular clock to a recent time for the mitochondrial ancestor of living humans. . ." (Ruvolo *et al.* 1994:8903; Ruvolo *et al.* 1993). Interpopulational heterogeneity is greater in mtDNA than in nDNA in modern humans; the mutation rate and probability of fixation is some ten times higher in the former. Thus, the mean numbers of nucleotide substitutions (D-loop region) between pairs of individuals within populations is 0.0469 (vs. 0.165 in chimpanzees), and that between populations is 0.0525 (vs. 0.140 in chimpanzees). Most human diversity (some 85%) is of course expressed polymorphically (within populations) rather than polytypically (between populations) (Lewontin 1972). Thus, p-demes are both informative and problematic from an actualistic perspective.

Modern Human Origins and the Pattern of Later Hominin Evolution

A number of estimates (Table 2, Appendix A) of the age of a modern human ancestor (*cf.* Long 1993) have been made, both from classical genetic markers (proteins, blood groups, etc.) and from mtDNA (either control region or coding sequences) lineage coalescence projections. The latter affords a measure of amount of evolutionary change along a lineage; extent of divergence between two lineages may also be evaluated (and is twice the former value). There is congruence in analysis of divergence according to several different analytical (statistical) procedures employed (MP, NJ-BS, UPGMA), although rooting of trees has sometimes afforded problems and attendant controversy. Recently additional evidence has been afforded by analysis of gene sequences of the (paternally-transmitted) Y chromosome (Pääbo 1995). All such estimates afford mean ancestral ages in a range between a hundred thousand to as much as half a million years ago, dependent on the time of ape (*Pan*) and hominin divergence. A maximum age would reflect genetic divergence rather than (usually subsequent) population divergence. A mean value between those extremes, of 250 ± 50 Ka, is as reasonable an approximation as might now be expected. African populations exhibit maximum sequence divergence, and more divergent mtDNA lineages (enhanced heterozygosity). There are contrary results as to the order of subsequent differentiation of European and Asian populations. Aboriginal Amerindian populations exhibit least mtDNA diversity, and that diversity comprises a subset of Asian mtDNAs, of markedly altered frequencies. All genetic evidence, from whatever source, unequivocally fails to project modern human roots into deep Pleistocene time. These include classical (protein) genetic markers, nDNA (RFLP) markers, microsatellite loci, Alu insertion (and STRP) polymorphisms, and, of course, mtDNA. Thus, there is a principal constraint against any phylogenetic formulation that posits prolonged anagenesis in the modern human (*Homo sapiens*) lineage and, correlatively, protracted histories for presumptive subspecies. This is wholly corroborated by the documented evolutionary history of other mammalian taxa, as, for example, Groves (1992) demonstrates.

Demes that make up populations, and populations themselves, are ephemera ("individuals and groups are like clouds in the sky or dust storms in the desert"; Dawkins 1989:34-35). Or, as Futuyma (1987:467) expressed it, "in the absence of speciation, much of the geographical variation we observe is ephemeral, leaving little imprint on evolution in the long term. This is chiefly because most local populations are ephemeral; over even moderately short spans of evolutionary time (tens or hundreds of thousands of years), the habitats to which populations are adapted shift, often over long distances, in consequence of climatic change. The most recent dramatic instance . . . is seen in the Pleistocene, during which the distribution of both temperate and tropical species changed markedly and repeatedly. . . ." In fact Futuyma's (1987:465) purpose was to "propose that because the spatial locations of habitats shift in time, extinction and interbreeding among local populations makes much of the geographic differentiation of populations ephemeral, whereas reproductive isolation confers sufficient permanance on morphological changes for them to be discerned in the fossil record."

In fact, are not subspecies — aggregates of geographically-delimited population clusters — similarly ephemeral? Subspecies are not units of evolution, have an arbitrariness in respect to their recognition and delineation, and are customarily considered of limited classificatory value, despite being accepted as "a genuine taxonomic category based on populations," when in fact "no nonarbitrary criterion is available to define the category subspecies nor is the subspecies a unit of evolution except where it happens to coincide with a geographic or other genetic isolate" (Mayr & Ashlock 1991).

A focus on p-demes and on their spatio-temporal linkages (or lack thereof) allows avoidance of the vagaries of subspecific attributions and enables a more immediate evaluation of the smaller units that participate in evolutionary processes. Such examination strongly suggests that such units, and as well the attendant larger subspecific groups of which they were evidently a part, were evolutionary ephemera. P-demic continuity is sometimes patently demonstrated, but such is far from always the case, and both punctuations and disjunctions/discontinuities are evidenced even among the limited (less than two dozen) demic instances recognizable after only five generations of research. Thus, even the near future promises to afford both further documentation of known demic units as well as evidence of those as yet wholly unknown.

Lahr (1992; also 1994) examined the assertion that sets of hominin cranio-dental features, considered by some as characteristic of the phyletic evolution of regional populations from local (archaic) antecedents, regularly exhibit higher incidences in such regions. In fact, whereas many features (of which 30 were extensively analyzed) "present a regional pattern, this pattern does not always correspond to that proposed by the model," and she found some even "occur in other populations with a higher frequency" (Lahr 1994). Thus, "these features do not support a multiregional origin, giving further support to the existing fossil, chronological and genetic evidence for a single African origin of all modern humans."

Waddle (1993, 1994) has recently examined, through matrix correlation tests, expected versus actual morphological distances, according to several phylogenetic models, between antecedent and succeedent hominin samples of mid- to late-Pleistocene age from Europe, western Asia, and parts of Africa. The basis of evaluation comprised 165 cranial characters/parameters, including discrete traits (52), angles (39), and metrics (74). Her "quantitative analyses . . . support a single origin for modern humans as opposed to continuous long term evolution within regions." She concluded that "models that hypothesize worldwide gene flow in opposition to evolutionary continuity explain observed fossil variation only if continuity is a relatively weak force compared with gene flow." Thus, her "results support the gradistic notion that coterminous taxa tend to be similar in morphology . . .," but "do not support the idea of evolutionary continuity within regions as indicated by the persistence of specific cranial features." This analysis, as many others before, "clearly refutes evolutionary continuity in Europe, and for the Neanderthals of southwest Asia." It demonstrates that a "single African origin model provides a better explanation for the cranial variation described in the [actual] morphological distance matrix . . ." (Waddle 1994). This conclusion is wholly in keeping with the substantial body of a diversity of genetic evidence pointing consistently in the same direction (Nei 1995).

Lieberman (1995:192) has stressed that the "choice of characters poses the greatest obstacle to resolving evolutionary relationships among human taxa." Characters must be rigorously evaluated and ultimately chosen on the basis of homology, demonstrable shared-derived polarity, and universality in occurrence across hypodigms, with attendant delineation and definition of character state(s) and consistency in scoring methods. It has been posited (Frayer *et al.* 1993:21-22) "that *combinations* of features with differing frequencies diagnose groups in the past, just as they are important in forensic diagnoses of racial identification." I concur, however, with Lieberman (1995:166) that this "argument that regional human clades are best characterized by *combinations* of derived and primitive characters rather than any *specific* derived characters is illogical," as in fact, minimally "only one real shared-derived character is needed to demonstrate common ancestry." Lieberman's useful examination of some 33 characters often employed in such studies, of which "30 features [were] proposed by Frayer *et al.* (1993) as regional shared-derived characters (synapomorphies) shows that most (87%) clearly do not support polycentric evolution [regional continuity] and many (27%) support the RA [recent African origin] hypothesis" (1995:176). Nonetheless much remains to be elucidated in respect to the developmental and ontogenetic aspects of characters and character complexes and the co-associations and interrelationships among them.

Some knowledge of essential demographic parameters is critical in the evaluation of models of hominin phylogenesis (Weiss & Maruyama 1976). However important, this is perhaps the single most unknown aspect of the hominin past, for which inference and speculation afford only limited insight. The foundation for such study was set out largely by F. A. Hassan (1981), and it is fair to say that there have been only limited advances subsequent to his remarkable effort, at least in respect to Pleistocene deep time. Moreover, parameters surely varied in the course of the Pleistocene, and thus inferences derived from extrapolations from the demographics of recent foraging peoples may well have only limited relevance and applicability to those most ancient hominin taxa within genus *Homo*. However, general principles and maximal parameter values may afford some basis for argumentation. Table 3 (see Appendix A) sets out some estimates of Pleistocene hominin population parameters.

The (maximal) vital rates, as estimated for late Pleistocene hunter/foragers, are as follows (Hassan 1975): (a) birth interval = 22 months; (b) live birth interval (allowing fetal death and sterility at 12%) = 27-30 months; (c) female reproductive span (allowing reproductive capacity — nubility — at 16 years, and adult female longevity equal to 29 years) = 13 years; (d) live births/adult female [allowing for 10% maternal mortality and above parameters b and c] = 4.7 offspring; (e) surviving offspring/adult female (allowing infant mortality 50%) = 2.35 offspring. Annual population growth rate (allowing generation span of 24 years and life expectancy (after 15) as 17-18 years) is estimated at 0.73%. Subsequently, Hassan (1981) suggested a downward revision of this value to 0.011%. Some quite different parameters may be expected for earlier time spans in the Pleistocene and among such different p-demic taxa. Overall, life span is short, there is a prolonged offspring-spacing interval, and a low rate of child survivorship prevails.

A series of principles emerge from such vital rate parameters, given the demographic structures of foraging peoples. Overall populations are very small to small. Isolates, composed of family units, bands (of about 25), and demes (of about 500), are small overall and semi-continuously or discontinuously (parapatrically) distributed spatially (broadly corresponding to an 'island' or 'neighborhood' model). Population densities are low. Disjunctive distributions between larger population aggregates are both effected by density factors and ratio of food yield to energy cost and constrained by paleoclimatic/paleoenvironmental changes which impact associated habitats. Populations are largely in a state of equilibrium, with low to extremely low (arithmetic) growth rates (r), largely a balance between fertility and mortality rates, and thus characterized by minimal growth in numbers and, correlatively, very slow geographic expansion. Following the values posited in Table 3, doubling time (Dt) approximates a million years (990 ky) in the earlier Pleistocene. This approximation might well seem excessively slow considering the consequences of range extension and population growth attendant upon expansion into new (Eurasian) latitudes and habitats. Intermediate and thus elevated values between this excessively slow rate and the accelerated growth in the Late Pleistocene surely must have prevailed between ~1.0-0.25 Ma. In the Middle Paleolithic, Dt is substantially accelerated (128 ky, 6400 generations), and in the Upper Paleolithic it rises markedly to 6.3 ky (minimally 310 generations, at 20 years/generation), or some 20 times faster.

The distribution of pair-wise genetic differences, exemplified by mtDNA mismatch distributions, are directly relevant to population size parameters (Slatkin & Hudson 1991; Rogers & Harpending 1992; but compare Marjoram & Donnelly 1994). These reflect the number of nucleotide site differences between each of any pair of individuals. Differences among sequences within a population are mismatch distributions; differences among sequences from different populations are intermatch distributions (Harpending *et al.* 1993). A smooth unimodal or wave-shaped curve of aggregate distribution reflects ancient population growth, and its position on a scale of number of sites exhibiting differences between samples reflects timing and magnitude of growth (the curve peak corresponding to the maximum concentration of coalescent events). The presence of substantial variability in mtDNA trees among modern humans is a reflection of a "relatively recent expansion in size" of such human populations (Rogers & Jorde 1995). Elucidation of such population growth is enabled by knowledge of mutation rate (u, estimated at 2% or 4%/10^6 years); female population size prior to (N_0) and consequent upon (N_1) expansion; time (t) in generations (since expansion) and time (T) since expansion in units of $1/(2u)$ generations. The several divergence rates afford population expansion ages between (T) 33-75 Ka (at 4%, $u = 1.5 \times 10^{-3}$), or 200 generations, and (T) 66-150 Ka (at 2%, $u = 7.5 \times 10^{-2}$), or 4250 generations. The N_0 is considered as $< 7 \times 10^3$ and N_1 as at least 1.50×10^5 (hence a breeding population of ca. 3×10^5), based on an assumption of random mating. If, on the other hand, populations were geographically structured, as seems quite likely, then initial population was only a fifth as large (some 1,500 females) and the post-growth population less (ca. hundred-fold).

The pattern of intermatch (i-m) distribution (between populations) may be broadly (or closely) coincident with a mismatch (m-m) distribution (within a population). Rogers & Jorde (1995; Rogers 1995) consider that "this pattern suggests that an expansion [in population] either preceded or coincided with the separation of these populations." However, and not uncommonly, an intermatch distribution may precede diagrammatically (= antecede temporally) a m-m distribution, suggestive of an intervening interval of 30 up to 50 ky. This situation is suggestive of either of two explanations, each depending upon population structure, gene flow, and subsequent growth: (a) an initial, broadly panmictic population splits, with weak or absent gene flow, producing the i-m distribution, subsequent to which bottlenecks (or expansions) result in differing m-m distributions within the derivative populations; or (b) sub-populations of a larger population, linked by only weak gene flow, are largely separate, producing an i-m distribution wave, and with mutual subsequent expansion derive distinctive m-m distribution. The essential point is that "there is good reason to believe that the major human populations separated long before the expansions that are reflected in the within-group waves" (Rogers & Jorde 1995:12). In any case, initial populations are small, some 1,500 (breeding) females, and are characterized by subsequent (substantial) growth, thus reflecting a bottleneck situation (see also Maynard Smith 1990; Wills 1990).

Ayala (1995) considers that "the theory of gene coalescence suggests that . . . human ancestral populations had an effective size of 100,000 individuals or greater," and that "molecular evolution data favor the African origin of modern humans, but the weight of the evidence is against a population bottleneck before their emergence." In his most recent evaluation Takahata (1995) asserts that "all studies of human mtDNAs . . . suggest that the coalescence time is shorter than 0.2 my (10^4 generations) definitely much shorter than 1 my (5×10^4 generations)," that "the mtDNA diversity in the current human population was generated during the late Pleistocene," and that population (N_e values) fluctuated in the Pleistocene such that bottlenecks were real and important, and "local human populations underwent frequent extinction/restoration because of increased dispersal and adverse environmental conditions in the Old World."

Pairwise comparisons and, in fact, "all available genetic evidence is consistent with the proposition that the major [modern] human populations separated from a small initial population roughly 100,000 years ago and that most of these separate populations experienced a bottleneck, or an episode of growth several tens of thousands of years later" (Rogers & Jorde 1995:32). Harpending *et al.* (1993:494; see also Harpending 1994a, 1994b) concluded that "our results show that [modern] human populations are derived from separate ancestral populations that were relatively isolated from each other before 50,000 years ago. Major population expansion took place between 80,000 and 30,000 years ago — 80,000 years ago in Africa and perhaps 40,000 years ago among the ancestors of the Europeans." And, "the existence of between-group differences far older than within-group differences implies that the late Pleistocene expansion of our species occurred separately in populations that had been isolated from each other for several tens of thousands of years" (Harpending *et al.* 1993:495). More specifically humankind appears to have expanded from an initial size of some 10,000 (breeding) individuals to, at the least, thirty times that number. This is far below several population estimates for the Late Pleistocene, which range between 7 and 20 times greater. It is posited that a still smaller source population comprised minimally 1000 breeding individuals. Distinctive, small human population aggregates (subspecies = 'races') existed prior to their individual growth expansions. These perspectives afford no support to a continuity (so-called Multiregional

Evolution or Regional Continuity) model of hominin evolution, but they are consistent with a variety of recent (and African) replacement models.

Given small overall population size and low population density it is necessary to consider the distribution and arrangement of populational groupings. This entails the relationships between local (or minimum) bands (~ 25 individuals) to those of larger aggregates (maximum bands or demes, connubium or dialectical tribe of authors) and their arrangement and spacing in habitats. Some associated parameters include the exploitative range and minimum equilibrium size (MES) in regard to mating networks. The 'catchment territory' (or home range) is based on area of economic exploitation in respect to (radial) walking distance from home base, the latter commonly being 10 km and the former ~ 314 km² (Hassan 1981). Areas will vary in extent substantially in relation to population density (thus, 164 km² for 0.15 individual/km² for Hadza, and 6450 km² for 0.01 individual/km² for Caribou Eskimo). An MES of band aggregates (maximum band of Wobst), which reflects the size of a sub-population that can constantly afford members of a population with mates upon reaching maturity, has been assessed as 175-475 individuals (= 7-19 minimum bands) (Wobst 1974, 1976), thus maximally approximating demic size. These might comprise a maximally effective hexagonal matrix, with variable number of tiers of minimum bands, or a more linear arrangement (with correlatively enhanced distance between groups). Hassan (1981) considered 'regional groups' to comprise upwards of a thousand individuals and Africa plus Eurasia as having some 6,000 groups in 'Upper' Paleolithic times, and only some 1,200 groups in 'Middle' Paleolithic times.

The overall population sizes and associated densities projected by several authors (Table 3) are consistent in stipulating remarkably small population sizes, small regional population aggregates, very low densities, and extremely slow rates of population growth, particularly so within the mid- and earlier spans of the Pleistocene. Such parameters, probably coupled with very limited rates of range extension into virgin domains, initially including unoccupied latitudes of Eurasia, must be adequately acknowledged in any models of later hominin evolution. If an extra-African dispersal occurred broadly coincident with the Olduvai (N) subchron (~ 2.0-1.75 Ma) — equivalent to the ultimate cold (boreal to tundra)-to-warm/temperate climatic cycle of the Netherlands (late) Tiglian chronostratigraphic stage — that event is coincident with the appearance and most of the span of African *Homo ergaster*, the Dmanisi occurrence (Georgian Caucasus), an artifactual situation at Erq el Ahmar formation (Jordan rift), and, at the younger end, perhaps ancient hominin penetration into the Sunda Shelf. At the older end this would be broadly coincident with an interval of enhanced aridity across equatorial African latitudes, as documented in oceanic and continental sedimentary records. Such an event might thus attest to dispersal, through range extension and population fission (demic budding), into the southern tier of Asia, prior to any hint of penetration into Europe proper (from which a number of appropriate fossil-bearing localities are known), or into middle latitudes of central and eastern Asia. The latter area witnesses such penetration at most half a million years later.

The single parameters set out (Hassan 1981:189; also 1980) for the source of such an event (though not so intended) would posit population size of some 400,000 (some 800 demes of 16,000 bands) and density of $0.020/km^2$; an inhabited area of some $20 \times 10^6 km^2$ is also posited. The latter value encompasses some two-thirds of the African continent and thus might appear inordinately high. Some approximate estimates (in $10^6 km^2$) of major recent African biotic subregions are Saharan (6), Sahelian (3), lowland tropical forest (2), and savanna/grasslands (15). The extent of the first and third subregions, in terminal Pliocene time, are perhaps most critical to the question of habitability and hence to estimates of population distribution and inferences of population density. Their extent changes in accordance with low deep-sea temperatures and enlarged global ice-volumes, and within the time in question would have been influenced by planetary obliquity forcing, with substantial attendant displacement southward of biozone boundaries. The dearth of direct archeological documentation, either at all or in some acceptable geochronological context, sorely limits explicit reliance on that resource. Nonetheless an overall initial distribution between 15°N and 30°S is reasonably approximate, but with demes and their inclusive regional groups, 400 in all by Hassan's estimate, concentrated about riverine and lacustrine basin hydrogeographic regimes, effectively tethered as hominins must be, and thus often of quite disparate and disjunctive distribution. Consequently, it is rather more likely that overall distribution (inhabited area) was only a half or so as large as posited, that total population was substantially less than posited (half as large, perhaps), and that populations were markedly discontinuously distributed due to various resource limitations.

Further, there is the question of human populations of later (mid-Pleistocene) antiquity and the relevance of demographic parameters to questions of modern human origins:

> Only when gene exchanges occur frequently among local populations and the total number of breeding individuals in the whole population is kept as small as about 10,000 does the multiregional hypothesis become compatible with the estimated age of the LCA [last common ancestor] . . . However, it is difficult to explain how such a small number of individuals could occupy vast areas of Africa and Eurasia over the last 1 myr while maintaining an evolutionary status as a single species. A more likely explanation is that the age of the LCA indicates that modern humans originated much less than 1 myr ago without integrating the substantially diverged *H. erectus* genes. This, together with the premise that the genetic diversity in the oldest parental population is greater, provides support for the recent African origin of modern humans. (Horai *et al.* 1995:536.)

It has been noted before that world population for pre-modern humans has been estimated (by Hassan 1981; also 1980), on the basis of a diversity of parameters drawn from hunter/gatherer analogues, as $1.0 \pm 0.2 \times 10^6$. There have been no real efforts to individualize size estimates, much less density evaluations, for Africa vs. Eurasia. As there are correlative environmental changes in subtropical and tropical latitudes, consequent upon changes in global ice volume and deep-sea temperature, particularly attendant upon eccentricity modulation since 0.9 Ma producing one hundred thousand year paleoclimate periodicity, it might be argued that Eurasian populations waned when African populations waxed and vice versa under the impress of such glacial/interglacial regimes. Any model of population size requires appropriate elaboration and incorporation of such parameters.

Various estimates have been proposed from the perspective of population genetics, with particular reference to mtDNA coalescence, mismatch (nucleotide substitution) differences, and simulation studies. Among the lowest such (restricted) values are those of total N (source population) of < 3,000 individuals and N_e female of 40-600 individuals (Sherry et al. 1994). Similarly, $N = 1.25 \times 10^5$ and N_e female of ~2.5 × 10^4 was posited by some of the same authors (Harpending et al. 1993). In their further elaboration of the same model, considering initial (N_0) and post-expansion (N_1) parameters, values of less than 7,000 (perhaps 5,000) N_0 female and 150,000 N_1 females (thus, N_e of ~ 300,000) were posited under random-mating (panmictic) circumstances. In a much more likely geographically-structured situation the respective values were N_0 females = ~ 1,500 and N_1 females = 150,000. Takahata (1991, 1993a, 1993b) considered that the effective (N_e) population size, since the lower Pleistocene, probably averaged 10^4 (perhaps 10^5, considering evidence from Mhc). However, if overall population is some 10^7 then the total N_n (number of breeding individuals/populational unit × number of sub-populational units) is considered as ~10^6 (Takahata 1994). There is reason to believe that there has been "frequent extinction and recolonization (turnover) of subpopulations in the lineage leading to modern humans," due to relatively small local human populational differentiation, and the values that must thereby obtain between posited N_e and values of N and n given total population approximating a million. (It is apparent that sub-populations with reduced N "are more liable to extinction," and also that such values will be affected by the viability of groups as reflected in n.) Such circumstances could well attend substantially enhanced deterioration of environmental conditions and their attendant effects on life ways, on intensive adaptations in locales to localized resources, and to restriction of gene flow between increasingly discontinuously distributed subpopulational units.

In conclusion, I again cite Takahata (1991:594) in respect to population parameters and gene flow among Pleistocene hominins, with particular respect to modern human origins:

> If . . . mutant genes, that might be responsible for the evolution of H. sapiens, were favored in any deme, they would spread over the entire population with high probability... However, the required time depends strongly on the interplay among various population parameters . . ., and if the multiregional hypothesis assumes a large number of demes and $Nm \le 0.1$, it is unreasonable to think that even such favorable mutations could spread over the entire human population during the Pleistocene. (N = effective size of each deme; m = per generation mutation rate)

Epilogue

On an empirical level, human evolutionary studies in recent years have shown substantial progress and enhanced vitality. They have been especially marked by greatly expanded documentation of the hominin fossil record, now to ~ 5 Ma, recovery of representatives of hitherto unknown taxa, and enlargement of various p-deme samples. In a number of instances, sophisticated in-depth study of various aspects of skeletal biology, including seminal studies of growth and maturation, have afforded requisite morphological profiles of newly recognized taxa and much expanded knowledge of variability and other parameters of evolutionary biological relevance in previously recognized, but insufficiently documented or studied, groups. The elaboration and application of a diversity of geochronological methods, some only in the last decades, have significantly impacted age assessments and correlations of local, provincial, and regional fossiliferous/archeological occurrences and successions. The development and refinement of the GPTS, in conjunction with such geochronological efforts, in both deep-sea and continental sedimentary records, has afforded a hitherto unenvisioned time scale for temporally circumscribed late Cenozoic global events. At more local, even locality-specific levels, there has been greatly enhanced concern with stratigraphic, micro-stratigraphic, pedological, and related studies relevant to elucidation of paleogeographic circumstances at increasingly refined levels of analysis, such that efforts at paleo-'landscape' recognition and study can be considered feasible and even fruitful. Taphonomy, the roots of which extend well back into the previous century, has expanded vigorously in keeping with these developments, and archeological studies have become an integral part of the emergent paleoanthropology endeavor. Paleoenvironmental investigations have increased remarkably in scope and in depth, in the perspective of newly formulated chrono-stratigraphic frameworks, and in relation to developments in paleoclimate analysis, simulation, and modeling and broad applications of palynological and isotopic analyses.

The study of modern human populations was increasingly encompassed within the span of population genetics upwards of forty years ago. However, the focus on traditional genetic markers was largely employed toward the purported elucidation of relationships between populations (especially sub-racial units), as well as aspects of adaptation and population structure and demographics; it was relatively strongly classificatory in many instances. There have been, of course, notable exceptions, and concerns with population histories in sub-recent time are repeatedly evidenced in the work of some human population geneticists, particularly L. L. Cavalli-Sforza and M. Nei, but including also those more immediately affiliated with biological anthropology. Moreover, a major impact has clearly been through investigations directed at the elucidation of phylogenetic relationships within primates, particularly within Hominoidea. Increasingly, nDNA and mtDNA have become the focus of such efforts, and this has led to renewed and redirected concerns with ascertainment of populational affinities, with branching relationships, and ultimately with the origins of modern humans (*Homo sapiens*). It is of some interest that, in the latter case, studies have been pursued very largely independently, rather than cooperatively and conjointly, by those investigators in population genetics/molecular biology and those in human paleontology. Joint authorship by such practitioners is almost unknown. This is not to deny that researchers have sought to follow and be current with the nature and consequences of others' research endeavors, although competences across this span are scarcely to be expected or even conceivable. The overt lack of cooperative endeavor is not exemplary, and the need for conjoint research and resultant publication is patently manifest. The burgeoning interest in the evolutionary history of humankind, from paleontological, behavioral, and genetic perspectives, demands closer affiliations between concerned investigators.

There would appear to be some gap or lack of congruence between developments of an empirical sort and their theoretical evaluation. This is, likely, a problem both ontological and epistemological. It has, surely, direct bearing on approaches to and perspectives on the evolutionary process within the hominin clade, and most especially conceptions of genus *Homo*, its origins and evolutionary history. It is probably true that an encompassing scenario of hominin evolution is beyond our grasp, now if not forever. Similarly, models and hypotheses of hominin diversification, including that of origins of genus *Homo* and of modern humans, are all surely incomplete, and some still more seriously flawed. All 'hard' formations of hominin phylogenesis are controversial and have been, and are being, subjected to severe criticism. The polycentric, so-called 'candelabra hypothesis' (*sensu* C. S. Coon) was effectively moribund upon its formal exposition, as was soon recognized by some (surprisingly not all) evolutionary biologists. The so-called 'Noah's ark' formulation has been presented in several guises (with catchy monikers), including 'Garden of Eden' and 'African Eve', and the less euphonic expressions of RAO theory (Recent African Origins), AHRM (African Hybridization/Replacement Model), and AOAM (African Origins/Assimilation Model), the latter two of which explicitly permit or claim some levels of inter-demic or higher level hybridization. All have their varied, usually few (if nonetheless vocal) adherents, and each has been somehow negatively evaluated by contrary opponents and several independent and more detached scholars (Aiello 1993; Lewin 1993).

An MRE (Multiregional Evolution, or Regional Continuity) model has been proposed (Wolpoff *et al.* 1984) and subsequently elaborated (Wolpoff 1989), purportedly as a null hypothesis subject to testing and falsification against such other models. It was originally put forward "as an attempt to show how the specific pattern in a polytypic species can be explained as a consequence of modern clinal theory" (Wolpoff *et al.* 1984:450). Its proponents sought to account, in genus *Homo*, for (a) morphological contrast between a central, source population and peripheral, derivative populations; (b) appearance early on of so-called 'regional continuity' features peripherally by comparison with the purported appearance, substantially later, of other such features centrally; and (c) the establishment and persistence of such presumed peripheral vs. central contrasts subsequently. This formulation grew, in part, out of a 'center and edge' hypothesis that reflected reduced (or minimal) polymorphism at the periphery of a species' distributional range such that, in the case of East Asian Pleistocene hominins, there were "peripheral populations with intraregional homogeneity and interregional heterogeneity" (Wolpoff *et al.* 1984:450). A trend toward peripheral monomorphism was taken to be a consequence of "continued drift in small populations, founder, and peninsula (bottleneck) effects," and with adaptive divergence a consequence of environmental plus random drift effects.

The MRE model is, in fact, as much or more a scenario as an explicit, testable hypothesis. It has sought to encompass pretty much all, without actually explaining much of anything. In any case, this scenario, at least applied to evolution within genus

Homo and modern human origins in particular, is seriously flawed, even in respect to its presumptive reference basis in the hominin fossil record. Since its more formal proposal and subsequent proclamation (*e.g.*, Frayer *et al.* 1993) in a consistently 'hard' form by a varying coterie of proponents, it has emerged as a fragile, frayed, and unfructuous conceptual framework. It is now largely bankrupt and can be, and has been, rejected on innumerable grounds, including various aspects of population genetics, demographics, paleogeography, archeology, and, indeed, the hominin fossil record itself. It has presumed, at least by implication, the highly questionable existence of a limited number of primeval regional populations, now considered by proponents as already *sapiens* and thus within which are to be found, perhaps incipiently, the sources (roots) of all modern human diversity. In this regard, I concur with Marks' perspective (1995:193) that:

> There is no evidence for a primordial division of the human species into a small number of genetic clusters that are different from one another. The fact is, we do not know how many basic groups of people there are, and it is very likely that there are *no* small number of groups into which a significant proportion of the biological diversity of the human species collapses.

MRE requires that natural selection drive African and African-derived hominin populations in Eurasia anagenetically and ineluctably toward the modern human condition. It has an almost omega-point inevitability about it. This is ultra-Darwinian in its most extreme sense (Eldredge 1995), and is difficult to conceptualize in relation to population aggregates, sizes, and distributions, gene flow in respect to migration/exchange rates, the role of transfigured landscapes and attendant habitat distribution, fragmentation, and uncongeniality under cyclic Pleistocene paleoclimatic mechanisms, not to mention inexplicably high selection pressures in situations strongly favorable to genetic drift. It has denied the role of speciation as a consequence of populational isolation, fragmentation, or diminution, usually rejected evidence of stasis in instances in which this effect is strongly indicated and hence of evolutionary interest and significance, and dismissed the probability and potential relevance of extinction events. Population dispersals, expansions, displacements, or replacements are seemingly unenvisioned or disallowed. Although provincial and regional evolutionary patterns and attendant processes are evidenced in hominin evolution, a 'hard' MRE hypothesis obscures rather than enlightens serious investigation into hominin phylogenesis.

The recent volumes of Howells (1993) and Tattersall (1993, 1995) reflect well and more fully the perspective and positions on hominin evolution that I have sought to adumbrate only briefly here.

Literature Cited

Aiello, L. C. 1993. The fossil evidence for modern human origins in Africa: A revised view. *American Anthropol.* 95:73-96.

Albrecht, G. H., & J. A. M. Miller. 1993. Geographic variation in primates: A review with implications for interpreting fossils. Pages 123-161 *in* W. H. Kimbel & L. B. Martin, eds., *Species, Species Concepts, and Primate Evolution*. Plenum Press, New York.

Andrews, P. J. 1992. Evolution and environment in the Hominoidea. *Nature* 360:641-646.

_____ , & D. B. Williams. 1973. The use of principal components analysis in physical anthropology. *American Jour. Phys. Anthropol.* 39:291-304.

Ayala, F. J. 1995. The myth of Eve: Molecular biology and human origins. *Science* 270:1930-1936.

Bailey, W. J. 1993. Hominoid trichotomy: A molecular overview. *Evol. Anthropol.* 2:100-108.

Bilsborough, A. 1973. A multivariate study of evolutionary change in the hominid cranial vault and some evolutionary rates. *Jour. Hum. Evol.* 2:387-403.

_____ . 1976. Patterns of evolution in Middle Pleistocene hominids. *Jour. Hum. Evol.* 5:423-439.

_____ . 1978. Some aspects of mosaic evolution in hominids. Pages 335-350 *in* D. J. Chivers & K. A. Joysey, eds., *Recent Advances in Primatology 3: Evolution.* Academic Press, London, UK.

_____ . 1983. The pattern of evolution within the genus *Homo. Prog. Anat.* 3:143-164.

_____ . 1984. Multivariate analysis and cranial diversity in Plio-Pleistocene hominids. Pages 351-375 *in* G. N. van Vark & W. W. Howells, eds., *Multivariate Statistical Methods in Physical Anthropology.* D. Reidel, Dordrecht.

_____ , & B. A. Wood. 1986. The origin and fate of *Homo erectus.* Pages 295-316 *in* B. Wood, L. Martin, & P. Andrews, eds., *Major Topics in Primate and Human Evolution.* Cambridge University Press, Cambridge, UK.

Birdsell, J. B. 1972. *Human Evolution.* Rand-McNally, Chicago, Illinois. 546 pp.

Bowler, P. J. 1986. *Theories of Human Evolution: A Century of Debate, 1844-1944.* Basil Blackwell, Oxford, UK. 318 pp.

Bräuer, G., & R. E. Leakey. 1986. The ES-11693 cranium from Eliye Springs, West Turkana, Kenya. *Jour. Hum. Evol.* 15:289-312.

Bräuer, G., & K. W. Rimbach. 1990. Late archaic and modern *Homo sapiens* from Europe, Africa, and southwest Asia: Craniometric comparisons and phylogenetic implications. *Jour. Hum. Evol.* 19:789-807.

Campbell, B. G. 1965. The nomenclature of the Hominidae, including a definitive list of named hominid taxa. *Occas. Pap. Roy. Anthropol. Inst., London* 22:1-33.

Cann, R. L., M. Stoneking, & A. C. Wilson. 1987. Mitochrondrial DNA and human evolution. *Nature* 325:31-36.

Carter, G. S. 1951. *Animal Evolution: A Study of Recent Views of its Causes.* Sidgewick & Jackson, London, UK. 368 pp.

Corruccini, R. S. 1974. Calvarial shape relationships between fossil hominids. *Yearb. Phys. Anthropol.* 18:89-109.

Cracraft, J. 1983. Species concepts and speciation analysis. *Curr. Ornithol.* 1:159-187.

_____ . 1989. Speciation and its ontology: The empirical consequences of alternative species concepts for understanding patterns and processes of differentiation. Pages 28-59 *in* D. Otte & J. A. Endler, eds., *Speciation and its Consequences.* Sinauer Associates, Sunderland, Massachusetts.

Dawkins, R. 1989. *The Selfish Gene.* Oxford University Press, Oxford. 352 pp.

Day, M. H., & C. B. Stringer. 1991. Les restes crâniens d'Omo Kibish et leur classification à l'intérieur du genre *Homo. Anthropologie* (Paris) 95:573-594.

Dewey, J. 1910. *The Influence of Darwinism on Philosophy, and Other Essays in Contemporary Thought.* Henry Holt, New York. 309 pp.

Dobzhansky, T. 1944. On species and races of living and fossil man. *American Jour. Phys. Anthropol.* (ns) 2:251-265.

Dorit, R. L., H. Akaski, & W. Gilbert. 1995. Absence of polymorphism at the ZFY locus on the human Y chromosome. *Science* 268:1183-1186.

Eldredge, N. 1995. *Reinventing Darwin: The Great Debate at the High Table af Evolutionary Biology*. John Wiley & Sons, New York. 244 pp.

_____, & J. Cracraft. 1980. *Phylogenetic Patterns and the Evolutionary Process*. Columbia University Press, New York. 349 pp.

Frayer, D. W., M. H. Wolpoff, A. G. Thorne, F. H. Smith, & G. Pope. 1993. Theories of modern human origins: The palaeontological test. *American Anthropol.* 95:14-50.

Futuyma, D. J. 1987. On the role of species in anagenesis. *American Nat.* 130:465-473.

Gabunia, L., & A. Vekua. 1995. A Plio-Pleistocene hominid from Dmanisi, East Georgia, Caucasus. *Nature* 373:509-512.

Gilmour, J. S. L., & J. W. Gregor. 1939. Demes: A suggested new terminology. *Nature* 144:333.

Goldstein, D. V., A. R. Linares, L. L. Cavalli-Sforza, & M. W. Feldman. 1995. Genetic absolute dating based on microsatellites and the origin of modern humans. *Proc. Nat. Acad. Sci. USA* 92:6723-6727.

Gould, S. J. 1988. On replacing the idea of progress with an operational notion of directionality. Pages 319-338 *in* M. H. Nitecki, ed., *Evolutionary Progress*. University of Chicago Press, Chicago, Illinois.

Groves, C. P. 1992. How old are subspecies? A tiger's eye-view of human evolution. *Archaeol. Oceania* 27:153-160.

Hammer, M. F. 1995. A recent common ancestry for human Y chromosomes. *Nature* 378:379-380.

Hapgood, P. J., & M. J. Walker. 1986. Analyse en composantes principales et classification hiérarchique de cranes du Pléistocène supérieur. *Anthropologie* (Paris) 90:555-566.

Harpending, H. C. 1994a. Gene frequencies, DNA sequences, and human origins. *Perspect. Biol. Med.* 37:384-394.

_____ . 1994b. Signature of ancient population growth in a low-resolution mitochondrial DNA mismatch distribution. *Hum. Biol.* 66:591-600.

_____, S. T. Sherry, A. R. Rogers, & M. Stoneking. 1993. The genetic structure of ancient human populations. *Curr. Anthropol.* 34:483-496.

Harrison, T. 1993. Cladistic concepts and the species problem in hominoid evolution. Pages 345-371 *in* W. H. Kimbel & L. B. Martin, eds., *Species, Species Concepts, and Primate Evolution*. Plenum Press, New York.

Hartwig-Scherer, S. 1993. Allometry in hominoids: A comparative study of skeletal growth trends. Universität Zürich, Inaugural-Dissertation (Philosophischen Fakultät II).

Hasegawa, M., & S. Horai. 1991. Time of the deepest root for polymorphism in human mitochondrial DNA. *Jour. Mol. Evol.* 32:37-42.

Hasegawa, M., A. Di Rienzo, T. D. Kocher, & A. C. Wilson. 1993. Toward a more accurate time scale for the human mitochrondrial DNA tree. *Jour. Mol. Evol.* 37:347-354.

Hassan, F. A. 1975. Determination of the size, density, and growth rate of hunting-gathering populations. Pages 27-52 *in* S. Polgar, ed., *Population, Ecology and Social Evolution*. Mouton, The Hague, The Netherlands.

_____ . 1980. The growth and regulation of human population in prehistoric times. Pages 305-319 *in* M. N. Cohen, R. S. Malpass, & H. G. Klein, eds., *Biosocial Mechanisms af Population Regulation*. Yale University Press, New Haven, Connecticut.

_____ . 1981. *Demographic Archaeology*. Academic Press, New York. 298 pp.

Henke, W. 1992a. Die Proto-Cromagnoiden-morphologische Affinitäten und phylogenetischen Rolle. *Anthropologie* (Brno) 30:125-138.

_____. 1992b. A comparative approach to the relationships of European and non-European Late Pleistocene and early Holocene populations. Pages 229-268 *in* M. Toussaint, ed., *Cinq Millions d'Années l'Aventure Humaine*. Études et Recherches Archéologiques de l'Université de Liège, no. 56, Liège, Belgium.

Horai, S., K. Hayasaja, R. Kondo, K. Tsugane, & N. Takahata. 1995. Recent African origin of modern humans revealed by complete sequences of hominoid mitochondrial DNAs. *Proc. Nat. Acad. Sci. USA* 92:532-536.

Howells, W. W. 1973. Cranial variation in man. *Pap. Peabody Mus. Archaeol. Ethnol. Harvard Univ.* 67:1-259.

_____. 1989. Skull shapes and the map. *Pap. Peabody Mus. Archaeol. Ethnol. Harvard Univ.* 79:1-189.

_____. 1993. *Getting Here: The Story of Human Evolution*. Compass Press, Washington, DC. 261 pp.

Hull, D. L. 1989. *The Metaphysics of Evolution*. State University of New York Press, Albany, New York. 331 pp.

Kidder, J. H., R. L. Jantz, & F. H. Smith. 1992. Defining modern humans: A multivariate approach. Pages 157-177 *in* G. Bräuer & F. H. Smith, eds., *Continuity or Replacement: Controversies in* Homo sapiens *Evolution*. A. A. Balkema, Rotterdam, The Netherlands.

Kramer, A. 1993. Human taxonomic diversity in the Pleistocene: Does *Homo erectus* represent multiple hominid species? *American Jour. Phys. Anthropol.* 91:161-171.

_____. 1994. A critical analysis of claims for the existence of Southeast Asian australopithecines. *Jour. Hum. Evol.* 26:3-21.

_____, & L. S. Koenigsberg. 1994. The phyletic positioning of Sangiran 6 as determined by multivariate analysis. *Cour. Forschungsinst. Senckenb.* 171:105-114.

Lahr, M. M. 1992. *The Origins of Modern Humans: A Test of the Multiregional Hypothesis*. Ph.D. Thesis, Wolfson College, University of Cambridge, Cambridge, UK. 399 pp.

_____. 1994. The multiregional model of modern human origins: A reassessment of its morphological basis. *Jour. Hum. Evol.* 26:23-56.

Landau, M. 1991. *Narratives of Human Evolution*. Yale University Press, New Haven, Connecticut. 202 pp.

Lewin, R. 1993. *The Origins of Modern Humans*. Scientific American Library, New York. 204 pp.

Lewontin, R. C. 1972. The apportionment of human diversity. Pages 381-398 *in* T. H. Dobzhansky, M. K. Hecht, & W. C. Steere, eds., *Evolutionary Biology, Vol. 6*. Appleton-Century-Crofts, New York.

Lieberman, D. E. 1995. Testing hypotheses about recent human evolution from skulls: Integrating morphology, function, development, and phylogeny. *Curr. Anthropol.* 36:159-197.

Long, J. C. 1993. Human molecular phylogenetics. *Annu. Rev. Anthropol.* 22:251-272.

Marjoram, P., & P. Donnelly. 1994. Pairwise comparisons of mitochondrial DNA sequences in subdivided populations and implications for early human evolution. *Genetics* 136:673-683.

Marks, J. 1995. *Human Biodiversity: Genes, Race and History*. Aldine de Gruyter, New York. 321 pp.

Maynard Smith, J. 1990. The Y of human relationships. *Nature* 344:591-592.

Mayr, E. 1950. Taxonomic categories in fossil hominids. *Cold Spring Harbor Symp. Quant. Biol.* 15:109-118.

_____, & P. D. Ashlock. 1991. *Principles of Systematic Zoology*. McGraw Hill, New York. 475 pp.

Mountain, J. L., & L. L. Cavalli-Sforza. 1994. Inference of human evolution through cladistic analysis of nuclear restriction polymorphisms. *Proc. Nat. Acad. Sci. USA* 91:6515-6519.

Nei, M. 1978. The theory of genetic distance and evolution of human races. *Japan Jour. Hum. Genet.* 23:341-369.

_____. 1982. Evolution of human races at the gene level. Pages 167-181 *in* B. Bonne-Tamir, ed., *Human Genetics, Part A: The Unfolding Genome*. Alan R. Liss, New York.

_____. 1992. Age of the common ancestor of human mitochondrial DNA. *Mol. Biol. Evol.* 9:1176-1178.

_____. 1995. Genetic support for the out-of-Africa theory of human evolution. *Proc. Nat. Acad. Sci. USA* 92:6720.

Nei, M., & A. K. Roychoudbury. 1974. Genic variation within and between the three major races of man, caucasoids, negroids, and mongoloids. *American Jour. Hum. Genet.* 26:421-443.

_____. 1982. Genetic relationship and evolution of human races. *Evol. Biol.* 14:1-59.

_____. 1993. Evolutionary relationships of human populations on a global scale. *Mol. Biol. Evol.* 10:927-943.

Pääbo, S. 1995. The Y chromosome and the origin of all of us (men). *Science* 268:1141-1142.

Pesole, G., E. Sbisá, G. Preparata, & C. Saccone. 1992. The evolution of the mitochondrial D-loop region and the origin of modern man. *Mol. Biol. Evol.* 9:587-598.

Planck, M. 1950. *Scientific Autobiography and Other Papers*. Williams & Wilkins, London. 192 pp.

Roebroeks, W. 1994. Updating the earliest occupation of Europe. *Curr. Anthropol.* 35:301-305.

_____, & T. van Kolfschoten. 1994. The earliest occupation of Europe: A short chronology. *Antiquity* 68:489-503.

Rogers, A. R. 1995. Genetic evidence for Pleistocene population explosion. *Evolution* 49:608-615.

_____, & H. Harpending. 1992. Population growth makes waves in the distribution of pairwise differences. *Mol. Biol. Evol.* 9:552-569.

_____, & L. B. Jorde. 1995. Genetic evidence on modern human origins. *Hum. Biol.* 67:1-36.

Ruvolo, M., S. Zehr, M. von Dornum, D. Pan, B. Chang, & J. Lin. 1993. Mitochondrial COII sequences and modern human origins. *Mol. Biol. Evol.* 10:1115-1135.

Ruvolo, M., D. Pan, S. Zehr, T. Goldberg, T. R. Disotell, & M. von Dornum. 1994. Gene trees and hominoid phylogeny. *Proc. Nat. Acad. Sci. USA* 91:8900-8904.

Salmon, W. C. 1990. *Four Decades of Scientific Explanation*. University of Minnesota Press, Minneapolis, Minnesota. 234 pp.

Schultz, A. 1936. Characters common to higher primates and characters specific to man. *Quart. Rev. Biol.* 11:259-283, 425-455.

Shea, B. T., S. R. Leigh, & C. P. Groves. 1993. Multivariate craniometric variation in chimpanzees: Implications for species identification. Pages 265-296 *in* W. H. Kimbel & L. B. Martin, eds., *Species, Species Concepts, and Primate Evolution*. Plenum Press, New York.

Sherry, S. T., A. R. Rogers, H. Harpending, H. Soodyall, T. Jenkins, & M. Stoneking. 1994. Mismatch distributions of mtDNA reveal recent human population expansions. *Hum. Biol.* 66:761-775.

Simmons, T. 1990. *Comparative Morphometrics of the Frontal Bone in Hominids: Implications for Models of Modern Human Origins*. Ph.D. Dissertation, University of Tennessee, Knoxville, Tennessee. 296 pp.

_____ , A. B. Falsetti, & F. H. Smith. 1991. Evolutionary patterns in the frontal bones of Pleistocene hominids from western Asia. *Jour. Hum. Evol.* 20:249-269.

_____ , & F. H. Smith. 1991. Human population relationships in the late Pleistocene. *Curr. Anthropol.* 32:623-624.

Simpson, G. G. 1961. *Principles of Animal Taxonomy.* Columbia University Press, New York. 247 pp.

_____ . 1963. The meaning of taxonomic statements. Pages 1-31 *in* S. L. Washburn, ed., *Classification and Human Evolution.* Aldine, Chicago, Illinois.

Slatkin, M., & R. R. Hudson. 1991. Pairwise comparisons of mitochondrial DNA sequences in stable and exponentially growing populations. *Genetics* 129:555-562.

Spuhler, J. N. 1985. *The Evolution of Apes and Humans: Genes, Molecules, Chromosomes, Anatomy, and Behavior.* Butler Memorial Lecture Series, University of Queensland. 21 pp.

_____ . 1988. Evolution of mitochondrial DNA in monkeys, apes, and humans. *Yearb. Phys. Anthropol.* 31:15-48.

_____ . 1989. Evolution of mitochondrial DNA in human and other organisms. *American Jour. Hum. Biol.* 1:509-528.

_____ . 1993. Population genetics and evolution in the genus *Homo* in the last two million years. Pages 262-297 *in* C. F. Sing & C. L. Hanis, eds., *Genetics of Cellular, Individual, Family, and Population Variability.* Oxford University Press, Oxford, UK.

Stoneking, M., S. T. Sherry, A. J. Redd, & L. Vigilant. 1993. New approaches to dating suggest a recent age for the human mitochondrial ancestor. Pages 84-103 *in* M. J. Aitken, C. B. Stringer, & P. A. Mellars, eds., *The Origins of Modern Humans and the Impact of Chronometric Dating.* Princeton University Press, Princeton, New Jersey.

Stringer, C. B. 1974. Population relationships of later Pleistocene hominids: A multivariate study of available crania. *Jour. Archaeol. Sci.* 1:317-342.

_____ . 1978. Some problems in Middle and Upper Pleistocene hominid relationships. Pages 395-418 *in* D. J. Chivers & K. A. Joysey, eds., *Recent Advances in Primatology 3: Evolution.* Academic Press, London, UK.

_____ . 1987. A numerical cladistic analysis for the genus *Homo. Jour. Hum. Evol.* 16:135-146.

_____ . 1992. Reconstructing recent human evolution. *Philos. Trans. R. Soc. London. B* 337:217-224.

_____ . 1994. Out of Africa – A personal history. Pages 149-172 *in* M. H. Nitecki & D. V. Nitecki, eds., *Origins of Anatomically Modern Humans.* Plenum Press, New York.

Swisher, C. C. III, G. H. Curtis, T. Jacob, A. G. Getty, A. Suprijo, & Widiasmoro. 1994. Age of the earliest known hominids in Java, Indonesia. *Science* 263:1118-1121.

Takahata, N. 1991. Genealogy of neutral genes and spreading of selected mutations in a geographically structured population. *Genetics* 129:585-595.

_____ . 1993a. Allelic genealogy and human evolution. *Mol. Biol. Evol.* 10:2-22.

_____ . 1993b. Evolutionary genetics of human paleo-populations. Pages 1-21 *in* N. Takahata & A. G. Clark, eds., *Mechanisms of Molecular Evolution: Introduction to Molecular Population Biology.* Japan Scientific Societies Press, Tokyo, Japan.

_____ . 1994. Repeated failures that led to the eventual success in human evolution. *Mol. Biol. Evol.* 11:803-805.

_____ . 1995. A genetic perspective on the origin and history of humans. *Annu. Rev. Ecol. Syst.* 26:343-372.

Tamura, K., & M. Nei. 1993. Estimation of the number of nucleotide substitutions in the control region of mitochondrial DNA in humans and chimpanzees. *Mol. Biol. Evol.* 10:512-526.

Tattersall, I. 1986. Species recognition in human paleontology. *Jour. Hum. Evol.* 15:165-175.
_____ . 1992. Species concepts and species identification in human evolution. *Jour. Hum. Evol.* 22:341-349.
_____ . 1993. *The Human Odyssey: Four Million Years of Human Evolution.* Prentice Hall, New York. 191 pp.
_____ . 1995. *The Fossil Trail: How We Know What We Think We Know About Human Evolution.* Oxford University Press, New York. 276 pp.
Tchernov, E. 1992. Dispersal — a suggestion for a common usage of this term. *Cour. Forschungsinst. Senckenb.* 153:21-25.
Templeton, A. R. 1993. The "Eve" hypothesis: A genetic critique and reanalysis. *American Anthropol.* 95:51-72.
Trinkaus, E. 1990. Cladistics and the hominid fossil record. *American Jour. Phys. Anthropol.* 83:1-11.
van Vark, G. N. 1984. On the determination of hominid affinities. Pages 323-349 *in* G. N. van Vark & W. W. Howells, eds., *Multivariate Statistical Methods in Physical Anthropology.* D. Reidel, Dordrecht, The Netherlands.
_____ . 1985. Some aspects of the reconstruction of human phylogeny with the aid of multivariate statistical methods. Pages 64-77 *in* K. C. Malhotra & A. Basu, eds., *Human Genetics and Adaptation*, volume I (Proceedings of the Indian Statistical Institute golden jubilee international conference). Indian Statistical Institute, Calcutta, India.
_____ , & A. Bilsborough. 1991. Human cranial variability past and present. *American Jour. Phys. Anthropol.* 95:89-93.
_____ , & W. Schaafsma. 1992. Advances in the quantitative analysis of skeletal morphology. Pages 225-257 *in* S. R. Saunders & M. A. Katzenbach, eds., *Skeletal Biology of Past Peoples.* Wiley-Liss, New York.
Vigilant, L., M. Stoneking, H. Harpending, K. Hawkes, & A. C. Wilson. 1991. African populations and the evolution of human mitochrondrial DNA. *Science* 253:1503-1507.
Waddle, D. M. 1993. *The Evolution of Modern Humans: Testing Models of Modern Human Origins Using Matrix Correlation Methods.* Ph.D. Thesis, State University of New York, Stony Brook, New York. 218 pp.
_____ . 1994. Matrix correlation tests support a single origin for modern humans. *Nature* 368:452-454.
Weiss, K. M. 1984. On the number of members of the genus *Homo* who have ever lived, and some evolutionary implications. *Hum. Biol.* 56:637-649.
_____ , & T. Maruyama. 1976. Archeology, population genetics and studies of human racial ancestry. *American Jour. Phys. Anthropol.* 44:31-50.
Whitfield, L. S., J. E. Sulston, & P. N. Goodfellow. 1995. Sequence variation of the human Y chromosome. *Nature* 378:379-380.
Wiley, E. O. 1978. The evolutionary species concept reconsidered. *Syst. Zool.* 27:17-26.
_____ . 1981. *Phylogenetics.* John Wiley, New York. 439 pp.
Wills, C. 1990. Population size bottleneck. *Nature* 348:398.
_____ . 1995. When did Eve live? An evolutionary detective story. *Evolution* 49:593-607.
Wobst, H. M. 1974. Boundary conditions for Paleolithic social systems: A simulation approach. *American Antiquity* 39:147-178.
_____ . 1976. Locational relationships in Paleolithic society. Pages 49-58 *in* R. H. Ward & K. M. Weiss, eds., *The Demographic Evolution of Human Populations.* Academic Press, London, UK.
Wolpoff, M. H. 1989. Multiregional evolution: The fossil alternative to Eden. Pages 62-108 *in* P. Mellars & C. Stringer, eds., *The Human Revolution: Behavioural and Biological*

Perspectives on the Origins of Modern Humans. Edinburgh University Press, Edinburgh, UK.

_____ , X. Wu, & A. G. Thorne. 1984. Modern *Homo sapiens* origins: A general theory of hominid evolution involving the fossil evidence from Asia. Pages 411-483 *in* F. H. Smith & F. Spencer, eds., *The Origins of Modern Humans*. Alan R. Liss, New York.

Wood, B. A. 1991. *Koobi Fora Research Project Volume 4: Hominid Cranial Remains*. Clarendon Press, Oxford, UK. 466 pp.

APPENDIX A

TABLES 1-3

<table>
<tr><td colspan="2" align="center">Key to Abbreviations of Table 1</td></tr>
<tr><td>CV</td><td>Canonical Variates Analysis</td></tr>
<tr><td>CVA</td><td>Canonical Variates Analysis, Q- or R-mode</td></tr>
<tr><td>DF</td><td>Discriminate Function Analysis</td></tr>
<tr><td>D/M</td><td>Darroch/Mosiman Shape Discrimination</td></tr>
<tr><td>D^2</td><td>Mahalanobis Distance Statistic</td></tr>
<tr><td>MC</td><td>Matrix Correlation</td></tr>
<tr><td>MRA</td><td>Multiple Regression Analysis</td></tr>
<tr><td>PAUP</td><td>Phylogenetic Analysis Using Parsimony</td></tr>
<tr><td>PC</td><td>Principal Components Analysis</td></tr>
<tr><td>P/s–sh</td><td>Penrose-size/shape Statistic</td></tr>
<tr><td>Rs</td><td>Row Standardization Method</td></tr>
<tr><td>UPGMA</td><td>Unweighted Pair-Group Method with Arithmetic Mean</td></tr>
<tr><td colspan="2">Group Integrity/Affinity</td></tr>
<tr><td>a</td><td>African</td></tr>
<tr><td>b</td><td>European</td></tr>
<tr><td>c</td><td>Levantine</td></tr>
<tr><td>d</td><td>Asian</td></tr>
<tr><td>e</td><td>Sundaland</td></tr>
<tr><td>f</td><td>Modern-like</td></tr>
<tr><td colspan="2">Nos. = Number of Specimens</td></tr>
</table>

TABLE 1. Multivariate Treatments of Hominin P-Demes by Various Authors

Author	Cranial Variables	Nario-kotome	Olduvai/ LLK	Tighenif	Kabwe	Irhoud	Florisbad	Omo-Kibish	Laetoli Ngaloba	KRM	Mauer/ Arago
1. Andrews & Williams (1973)	25							1			
2. Bilsborough (1973, 1976, 1978, 1983, 1984)	107	2	1		1						
3. Bilsborough & Wood (1986)	22	3	1								
4. Bräuer & Rimbach (1990), Bräuer & Leakey (1986)	44							2	1		
5. Corruccini (1974)	51/27		1		2	3	1	1			
6. Day & Stringer (1991)	13				1	1		2			
7. Hapgood & Walker (1986)	22				1						
8. Henke (1992a, 1992b)	32										
9. Howells (1989)	57				1	1					
10. Kidder et al. (1992)	17					1					
11. Lahr (1992)	23/30										
12. Simmons (1990)	11				4	3	1	2	2		
13. Simmons & Smith (1991)	11				3	3	1	2	1		
14. Simmons et al. (1991)	8/12					1					
15. Stringer (1974)	42				1	1		2			
16. Stringer (1978)	21			1	2	2		2			1
17. Stringer (1987)	11	2	1		2	2	2	1	1		1
18. Stringer (1992)	38					2	1	1	1		
19. Stringer (1994)	39					2	2	1	1		
20. Stringer (this volume)	39/35	2	1		4	2	1	1	1		1
21. van Vark (1984)	17				1	1		1			
22. van Vark (1985)	17				1	2					
23. van Vark & Bilsborough (1991)	19										
24. van Vark & Schaafsma (1992)	26										
25. Waddle (1993, 1994)	113/52				4	3	2	1	1		
26. Wood (1991)	19/7/16	17	2/4								1

TABLE 1 (*continued*). Multivariate Treatments of Hominin P-Demes by Various Authors

Author	Petralona/ Atapuerca	Neandertal	Skhul/ Qafzeh	Yunxian	Zhou- koudian	Dali	Jinniu- shan	Maba/ Hathnora	Brangkal	Glaga- homba	Trinil/ Sangiran	Ngandong	Procedures
1.	2b/d	6b/d	2b/f		5d							5e/d	PC
2.	2	10	2		5				1		1	4	PC/CVA-Q mode
3.		8			5								CVA-Q mode
4.		8	4										PC
5.	3	18	4		5				1		4	6	PC
6.	1	16	1		5							5	D²
7.	2	8	4		2						1	5	PC/CA
8.			5										PC/DF/MRA
9.		2	2f										PC/D²/Q-R-Mode
10.	2	11	4f										D/M shape; D²
11.				x	x	x	x	x	x		x	x	MRA/DF
12.		10	6f										PC
13.		11	5f										PC/UPGMA
14.		4	6f										PC/UPGMA
15.	3	13	1f		x						x	x	D²/CVA
16.	3	18	5f		x						x	x	P/s-sh
17.	1		4f		x						x		PAUP
18.		13	5f			1		1					P/s-sh/RS
19.		13	5f										P/s-sh
20.	5	15	5f										P/s-sh/RS
21.	2	12	3		2						1	4	PC/D²
22.		8	2		2							4	PC/D²
23.		2	3										D²
24.			2f										PC/D²
25.	2	13	5f										MC/Mantel-r
26.					2						3		CV/CA/D²

TABLE 2. Various estimates of the age of the common ancestor(s) of modern human populations from genetic evidence. A = Classic genetic markers (proteins, blood group loci); B = mtDNA; C = Y-chromosome. P/H = $Pan/Homo$ time of divergence.

Reference	Age of Divergence (Ka)	Groups/Conditions
A.		
Nei and Roychoudbury (1974)	~ 115-120	
Nei (1978)	~ 120	African vs. European
	~ 60	European vs. Asian
Nei and Roychoudbury (1982)	~ 120 (110 ± 34)	African vs. European/Asian
	~ 50 (41 ± 15)	European vs. Asian
Nei and Roychoudbury (1993)	100-200	African vs. non-African
	50-70	European vs. all non-African
	40-50	Australasian vs. Greater East Asian
	12-40	Native American vs. all Greater East Asian
Goldstein et al. (1995)	~ 156 (75-287)	Microsatellite polymorphisms, as measure of deepest human branching time
B.		
Nei (1982)	53-106	African vs. European
	45-90	Asian vs. African
	20-40	Asian vs. European
Cann et al. (1987)	~ 140-290	
Hasegawa and Horai (1991)	~ 280 ± 50	If P/H = 4 Ma
Vigilant et al. (1991)	~ 166-249	If P/H = 4-6 Ma
Nei (1992)	~ 207	If P/H = 4-6 Ma
Pesole et al. (1992)	~ 400	If P/H = 6-8 Ma
Stoneking et al. (1993)	63-356	95% confidence interval; if Papua New Guinea colonization ≤ 60 Ka
Templeton (1993)	~ 170	If P/H = 4 Ma
	~ 256	If P/H = 6 Ma
Tamura and Nei (1993)	~ 160	If P/H = 4-6 Ma

TABLE 2 (*continued*). Various estimates of the age of the common ancestor(s) of modern human populations from genetic evidence. A = Classic genetic markers (proteins, blood group loci); B = mtDNA; C = Y chromosome. *P/H* = *Pan/Homo* time of divergence.

Reference	Age of Divergence (Ka)	Groups/Conditions
B. (*continued*)		
Hasegawa *et al.* (1993)	~ 211 ± 111	Control region, sequence I
	~ 387 ± 84	Control region, entire sequence
Ruvolo *et al.* (1993)	~ 195	If *P/H* = 4 Ma
	~ 298	If *P/H* = 6 Ma
Mountain and Cavalli-Sforza (1994)	~ 100	African vs. non-African
	~ 700	"Average time since mutation of the ancestral alleles producing the current set of polymorphisms"
Horai *et al.* (1995)	143 ± 18	Last common ancestor of modern human mtDNAs
	70 ± 13	European vs. Asian, if *P/H* = 4.9 Ma
Wills (1995)	~ 436-806	Based on rate of accumulation of transversions, and *P/H* = 4.0-7.4 Ma
C.		
Dorit *et al.* (1995)	~ 270	Deepest human branching time, based on hominoid mutation rate
Hammer (1995)	188 ± 94	Deepest human branching time (coalescence) of Alu/YAP gene
Whitfield *et al.* (1995)	37-49	Minimum/Maximum age of branching (coalescence) of SRY gene

TABLE 3. Some Estimated Pleistocene Hominin Population Parameters

	Group Numbers (×10³)		Population Density (Ind./km²)	World Population (×10⁶)	Populated World Area (×10⁶ km²)	Population Growth Rate (r)	Doubling Time (Dt)(k.y.)
	Bands	Demes					
Upper Paleolithic							
Hassan (1981)	240	12	0.100	6.0	60.0	0.011	6.3
(6000 regional groups @ 1000/group)							
Birdsell (1972)			0.039	~2.2	57.5		
Weiss (1984)			0.050				
Harpending et al. (1993)			0.050	(~0.30)			
Middle Paleolithic							
Hassan (1981)	48	2.4	0.030	1.2	40.0	0.0054	128.0
(1200 regional groups @ 1000/group)							
Birdsell (1972)			0.032	~1.0	38.3		
Weiss (1984)	(5)		0.038	1.3	34.5		
Harpending et al (1993)				0.125	25-28		
Lower Paleolithic							
Hassan (1981)	32	1.6	0.025	0.8	30.0	0.00007	990.0
Birdsell (1972)			0.015	~0.4	27.0		
Weiss (1984)			0.038	0.5-1.3	34.5		
Harpending et al. (1993)			0.005		25-28		
Basal Pleistocene (Oldowan)							
Hassan (1981)	16	0.8	0.020	0.4	20.0		

Paleoanthropology and Preconception

Ian Tattersall
Department of Anthropology
American Museum of Natural History
Central Park West at 79th Street
New York, NY 10024

Although discussion of the human fossil record began before an evolutionary framework existed within which to understand it, certain canons of belief established during this early period remain with us to this day. Principal among these is a willingness to accept a maximum of morphological variety within *Homo sapiens* while at the same time minimizing diversity among related fossil species. The adoption of Darwinian concepts added to this unfortunate legacy the transformationist mindset which has since dominated paleoanthropology almost to the exclusion of taxic diversification. This preoccupation is in turn related to a general denial of the view of species as essentially stable entities, which has more recently expressed itself in a widespread rejection of the notion of punctuated equilibria. The resulting intellectual climate has encouraged the view that human phylogeny is essentially a matter of discovery rather than a pattern of relationships among taxa that requires analysis. In consequence much of hominin systematics has been deprived of a coherent theoretical basis. It is critical that we develop a clear appreciation of how mindset has influenced paleoanthropological belief, and that we adopt more realistic criteria for species recognition in the human fossil record.

"If I have seen farther than other men," said Isaac Newton, "it is because I have stood on the shoulders of giants." In these gracious words he acknowledged a debt that is universal in science, no less in these days than in his own. Every branch of modern science is the beneficiary not only of a great body of accumulated knowledge but of a perspective on that knowledge that has been laboriously achieved over a great period of time, and through the efforts of many gifted individuals. I hope that nothing I say here will be taken as in any way belittling the contributions to our science of our precursors in the field of paleoanthropology. I do want to point out, however, that the precious legacy of the past also brings with it a burden that we could well do without: not all received wisdom is truly worthy of that name. It is the task of every generation of scientists to sort through its intellectual legacy, and in doing so to retain and build upon what has stood the test of observation and repeated experience, while discarding the rest. Despite this self-evident obligation, I fear that over the long haul paleoanthropologists have proven less ready than scientists in many other fields of systematics to re-examine their assumptions.

We have paid a considerable price for this reluctance to question received wisdom. It has caused us, for example, seriously to neglect or to underestimate the taxic diversity that has arisen in the evolution of the human family (or, more strictly speaking, of the tribe Hominini). The historical roots of this tendency are clear. When the significance of the initial fossils of the first-known extinct human species, *Homo neanderthalensis*, began to be disputed by anthropologists and human anatomists in the late 1850s, there was no context other than that provided by living *Homo sapiens* within which this remarkable new phenomenon was to be understood. All of the living species of our closest (if quite distant) living relatives, the great apes, were known by that time, and it was pretty evident to all concerned that the big-brained specimen from the cave in the Neander Valley was no ape, even in the rather vague sense in which that term was then understood. As the Neanderthal debate commenced, a couple of years before the publication of *The Origin of Species*, there was no coherent conception that we are linked to our nearest living relatives by a series of extinct species that represent an evolutionary genealogy, a genealogy that does indeed tie us in with the apes, with the monkeys, with the primates as a whole, and ultimately with the entire diverse spectrum of living things.

It is hardly surprising, then, that the early European anthropologists and anatomists, who labored so hard to explain what the big-brained, beetle-browed, low- and heavy-skulled Neanderthal specimen represented, sought to place this strange phenomenon in the context of the world that they knew. Two principal possibilities presented themselves. The first of these was that the Neanderthaler was somehow pathological, the remains of a modern human victim of some obscure but morphology-modifying disease. The second was that these remains had belonged to a member of an extinct and primitive tribe, unrepresented in the modern world and thus unknown to science in its living form (although such peoples had long been celebrated in Western myths dating back to Classical times). Of all of the savants who commented on the Neanderthal find, virtually only William King (1864) was prepared to entertain the view that here was evidence of a separate, extinct species comparable in some ways to the apes as well as to humans.

Eventually, scientists were forced to discount the first of these possibilities, that of pathology, by repetitive discoveries of similar fossils in widely separated parts of Europe. Pathology simply could not explain the consistency of all of the accumulating Neanderthal finds. But the second alternative remained, and remains with us still, albeit in modified form. To this very day, the most egregious example of the ancient tradition of cramming a bizarre array of forms into the species *Homo sapiens* is provided by the allocation of the Neanderthals to the subspecies *Homo sapiens neanderthalensis*. The routine parading out of the Neanderthals in textbooks of physical anthropology as representatives of an extinct or reabsorbed subspecies of our own species thus continues to infect the mindsets of successive generations of physical anthropologists.

Perhaps ironically, this mindset was reinforced by the triumph of the evolutionary synthesis of the 1930s and 1940s (*e.g.*, Dobzhansky 1937; Mayr 1942; Simpson 1944) which among other things emphasized the importance of variation within species. For systematists accustomed to dealing with the variety of species in the living world (most of which possess much closer extant relatives than we do) this has been a salutary development, combating the ancient and perfidious typological tendency of one specimen, one name. But to anatomists and anthropologists, whose training and experience were traditionally limited to the documentation of variation within the one species *Homo sapiens* (with perhaps an occasional glance at the other hominoid genera, all notably unspeciose), it has proven dangerous indeed. It simply reinforced an already deeply entrenched viewpoint that was prepared to accept maximum variety within the one living species around which the sciences of human anatomy and anthropology revolve, while recognizing minimum variety at the species level and above in the fossil record of human precursors.

This viewpoint is still very much with us today and represents the aspect of received wisdom in paleoanthropology that cries out, above all else, for re-evaluation. Species provide the essential basis for all systematic analysis, whether of humans or of any other group. Yet our ideas of species in paleoanthropology still tend to be dominated by Charles Darwin's brilliant response to a specific set of beliefs that dominated Victorian society a century and a half ago. Darwin's idea, that species gradually transform themselves out of existence over vast stretches of time, was born of an essentially political need to combat the Biblically-grounded view of species as immutable. The proposition that gradual natural selection is the agent of such change gave this proposal a powerful appeal. Its attractiveness was further strengthened by the addition of a genetic basis when the architects of the "New Synthesis" argued that virtually all evolutionary phenomena could be reduced to the influence of natural selection working on the gene pools of populations over vast periods of time.

It has to be pointed out, however, that this putatively all-embracing mechanism ignores the contingent effects of climate and geography. These crucial forces are largely random with respect to adaptation, and they are also decidedly irregular with respect to time. What's more, the gradualist mechanism also disregards the little that we know about speciation in mammal populations and has distorted our interpretations of the fossil record. It does, though, have the advantage of a magnificent simplicity and of the intellectual appeal that simplicity always exerts on the human mind, a mechanism that seems to crave elegant explanation. The world may in fact be a rather complicated, unfathomable place; concordantly complex explanations, however, seem to be singularly unattractive to our ratiocinative faculties.

For better or for worse, though, if we wish to discover pattern in the fossil record we need to have a realistic working conception of what we mean by species. I use the modifier ("working" conception) here for two reasons. First of all, we know that speciation (in the sense of the splitting of lineages rather than their gradual transformation) must occur, for otherwise life could never have diversified. However, we know little about the mechanisms that underlie this fundamental process. Indeed, it seems likely that speciation is not in fact a genetic event of a specific kind, but that it results instead from genetic disruptions of a particular class, underlain by a variety of mechanisms (Godfrey & Marks 1991; Tattersall 1992). I have summarized elsewhere (Tattersall 1992) some of the problems that are posed by the various major species concepts currently on offer. Suffice it to note here that as long as we do not fully understand what speciation is, we will find it hard to be definitive about what species are.

On the other hand, it is already clear what species are not: they are not simply the passive results of accumulating adaptive change (Tattersall 1991). This is dramatically underlined by the recent study of Sturmbauer & Meyer (1992), who found genetic, morphological, and taxic divergence to be effectively unrelated among species of cichlid fishes living in various lakes of the East African Rift system. The obvious problem for paleontologists is that, if there is no close correlation between speciation, time, and any specifiable degree of morphological change in the skeleton and dentition (the only body systems for which information on extinct forms is routinely available), there can be, for the time being at least, no objective means of recognizing species in the fossil record. There exists, of course, a vague association between morphology and speciation in the sense that speciation has to intervene in the context of a geographically differentiating population, but there is no direct link between the two.

I have suggested elsewhere (Tattersall 1986, 1991, 1992) that this should not be a major problem in practical terms if we are prepared to realize that species in general represent a rather low level of morphological differentiation. So low, indeed, that where distinct skeletodental morphs can be recognized we are on pretty firm ground in recognizing entities which are at the very least species in a genetic sense. Indeed, such morphs are perhaps just as likely to represent monophyletic groups of closely related species as they are to represent individual species. When compared to conventional interpretations of the human fossil record (at most a mere five species between *Australopithecus afarensis* and *Homo sapiens*), this observation suggests that paleoanthropologists have traditionally been prone to overestimate how much morphological variety is likely to accumulate within a species before it becomes extinct or splits into two or more daughter species. In part this tendency has resulted from the historical circumstances already adumbrated, but it has clearly been reinforced by another related factor as well. Several years ago Eldredge (1979) pointed out that the evolutionary process possesses two aspects: morphological transformation and taxic diversification. To transformationists, evolution centers around accumulating changes in genes and their products, and the role of the paleontologist is to track such change through time. The focus is thus upon aspects of morphology and the potential agents of selection bearing upon them. How morphology is packaged into species is very much a subsidiary concern, if indeed it is a concern at all.

The tradition in paleoanthropology has thus overwhelmingly been one of transformationism. The result has been that in our science the second major aspect of evolution, the splitting of species and hence the origin of taxic diversity, has tended to be relegated to a secondary role, to be ignored, or even to be totally denied. In witness to this one has only to think of the single-species hypothesis that enjoyed such a vogue in the 1960s and early 1970s (*e.g.*, Wolpoff 1973) or of its precursor and latter-day descendant, the movement to sink *Homo erectus* as well as all subsequent hominins into the species *Homo sapiens* (*e.g.*, Thoma 1970; Wolpoff 1992). This transformationist tradition is closely linked, of course, with the pervasive notion in paleoanthropology that the unraveling of human evolutionary history is effectively a matter of discovery (Eldredge & Tattersall 1975): if we crawl over enough outcrops and discover enough fossils, all will eventually be revealed unto us. "We need more fossils" is an obligate part of the paleoanthropological litany, while the lament that "we need better methods of phylogenetic analysis" is far more rarely found in the concluding sections of papers on the human fossil record.

Perhaps I may be permitted a personal reminiscence here. As a graduate student I watched fascinated as visiting researchers diligently made reams of notes on fossils in the collections room where I had my desk. I had a hard time figuring out what they were actually doing, even though I was presumably there to learn to emulate them. Eventually I plucked up the courage to ask a distinguished paleoanthropologist to reveal the secret of how to study fossils. He thought for a moment, then said: "Well, you look at them for long enough, and they speak to you." Of course, I have since come to learn that there is actually a great deal more wisdom in that answer than I realized at the time, but it does illustrate very neatly how, right up through the 1960s, the practice of paleoanthropology, like that of vertebrate paleontology in general, was very much an intuitive affair. The reconstruction of phylogeny was bereft of any rigorous theoretical framework and in general did not yield hypotheses that were objectively comparable one with another. The field was discovery-driven. Admittedly, this had brought paleontology a remarkably long way, but it is no wonder that I was delighted to make my own discovery of cladism after arriving at the American Museum of Natural History.

My pleasure at discovering this theoretically coherent, practical approach to the analysis of relationships among fossil and living species has not, however, been as widely shared as it might have by my colleagues in paleoanthropology. Despite the practical difficulties that cladism undoubtedly does present in such areas as the definition of characters and the determination of morphocline polarities, it is, I think, principally as a result of the long-standing transformationist bias in paleoanthropology that the cladistic approach to the reconstruction of phylogeny has made so little impression upon the field. The fact that much of the vocabulary of cladism has been co-opted into paleoanthropological jargon should not be mistaken for wholesale adoption of its principles. While it is probably true by now that most vertebrate paleontologists outside the primate field have come to recognize the value of proceeding from simple hypotheses of sister relationships to more complex hypotheses of ancestry and descent, and only finally adding to the brew such ingredients as ecology and adaptation (Tattersall & Eldredge 1977), paleoanthropological hypotheses still tend to be introduced at the level of the scenario (a complex mishmash of relationships, time, adaptation, and ecology) rather than at the simpler levels at which hypotheses are actually (or at least potentially) testable. It is also because of the weight of transformational traditionalism that many paleoanthropologists have been so ready to reject notions of punctuated equilibrium as opposed to phyletic gradualism in evolution. Such out-of-hand rejection has been easier to accomplish in paleoanthropology than in most branches of vertebrate paleontology. This is for the simple reason that we are dealing in our field with a single extremely closely related group of organisms. Within such close-knit groups, "intermediates" between members — traditionally accepted as *prima facie* evidence of continuity — are particularly easy to find.

This is so for two reasons, having to do both with homology and homoplasy (the independent acquisition of evolutionary novelties in two or more species). The more closely two species (or populations) are related, the more alike they will almost certainly be genetically. They are likely to overlap substantially or totally in the ranges of variation of almost all of their morphological characters. Further, the more similar such populations are genetically, the more likely it becomes that the same genetic innovations will arise independently within each. The practical problems that these likelihoods raise for the working paleontologist become yet more acute when we factor in the consideration that closely related species will not only share the same major constellation of adaptive characters, but that they will generally experience at least broadly equivalent ecological pressures. Natural selection takes place at the level of the local population, and in similar circumstances closely related populations are likely to respond to ecological pressures or other agents of natural selection upon them in similar ways. These various considerations will hold true even when such local populations have become individual evolutionary entities. When, that is, speciation (which takes place for reasons that are not necessarily or even at all related to adaptation) has intervened between them.

This is a problem that has received remarkably little attention from paleoanthropologists. For all the interminable measuring that has been done, as far as I am aware few people have troubled to investigate the patterns of distribution of skeletodental characteristics between defined populations of living primates at the specific and intraspecific levels. When Schwartz and I (Schwartz & Tattersall 1991) tried this recently on a group of extant lemur species, using the quantitative parsimony procedure PAUP on what we had thought was a very substantial character set, we came up with 80 alternative trees of equal probability. This indeterminate result was evidently caused by extensive homoplasy (as might have been expected, for the reasons already given). It further resulted in the conclusion (Tattersall 1993) that the number of taxa in our sample, at both the specific and the infraspecific levels, would have been seriously underestimated by a paleontologist using our data set to sort out the alpha taxonomy of this closely related group in ignorance of the "biological" information (behavioral, distributional, external morphological, karyotypic) which we had possessed at the outset of our study.

We have no reason to believe that the group we chose for our study is in any way atypical for primates in general, and our experience dramatically illustrates the difficulties that lie in wait for systematists, such as paleoanthropologists interested in the genus *Homo* or even in the tribe Hominini, who are concerned with taxa that are differentiated at a very low level. The empirically demonstrated as well as theoretically expected tendency towards high levels of homoplasy among closely related species points towards a built-in likelihood that we will underestimate the abundance of species in the human fossil record, and it also suggests that we should not be surprised to encounter supposedly "Neanderthal-like" characters in early *Homo sapiens* from Europe. Such characters, particularly in the supraorbital and occipital regions, were cited in earlier years in support of a Neanderthal ancestry for modern humans in that part of the world, and more latterly have been adduced as evidence for some admixture between the two.

To return to matters of mindset, what you expect of the evolutionary process clearly affects what you will see in the fossil record. The ongoing debate between proponents of the "multiregional" and "center of origin" models of the emergence of *Homo sapiens* reflects theoretical expectations at least as much as it does the facts of the fossil record. Those who, in the spirit of the "New Synthesis," see human evolution as a long trudge from benightedness to enlightenment will grasp at the "evidence" for continuity furnished by the expectable homoplasy between closely related hominin species. Those who, on the other hand, prefer to acknowledge the largely contingent nature of the real world will look for centers of origin. For little as we know about the process, or processes, of speciation among primates, we do know that it, or they, do require the physical division of a pre-existing species and the interruption of free-flowing genetic contact between its components. The isolation of infraspecific populations is thus a prerequisite for speciation, and the occasions for such isolation can rarely have occurred more frequently than during the dramatic climatic and glacio-eustatic fluctuations of the Pleistocene, the epoch which saw the appearance of most of the innovations documented in the human fossil and archeological records.

My purpose here, however, is not to espouse any specific views of how human evolution proceeded at any particular point in time. That will be done much more competently than I can by the other contributors to this symposium. I want, rather, to emphasize that at its fundamental systematic level our study of human evolution is in essence a search for historical pattern, and that it is thus a matter of analysis as much as it is of discovery. Fossils are essential, but they don't speak. Particularly at the low levels of morphological differentiation involved in paleoanthropology, the patterns we perceive are as likely to result from our unconscious mindsets as from the evidence itself. This, above all, is why I believe that it is our responsibility to our illustrious predecessors, as much as it is to ourselves, to re-examine our received attitudes from time to time, to ensure that we are building on the foundations of the past, rather than staying entombed in their ruins.

Acknowledgements

For the opportunity to make these remarks I should like to acknowledge my gratitude to the Paul L. and Phyllis Wattis Foundation Endowment and to the California Academy of Sciences. Appreciation for the organization of an exceptionally stimulating symposium is also due to Roy Eisenhardt, the Academy's Executive

Director Emeritus, to Linda S. Cordell, former Irvine Curator of Anthropology, and above all to Deborah W. Stratmann, the Symposium Coordinator.

Literature Cited

Dobzhansky, T. 1937. *Genetics and the Origin of Species*. Columbia University Press, New York. 364 pp.

Eldredge, N. 1979. Alternative approaches to evolutionary theory. *Bull. Carnegie Mus. Nat. Hist.* 13:7-19.

_____, & I. Tattersall. 1975. Evolutionary models, phylogenetic reconstruction, and another look at hominid phylogeny. Pages 218-243 *in* F. S. Szalay, ed., *Contributions to Primatology, Vol. 5: Approaches to Primate Paleobiology*. Karger, Basel, Switzerland.

Godfrey, L. R., & J. Marks. 1991. The nature and origins of primate species. *Yearb. Phys. Anthropol.* 34:39-68.

King, W. 1864. The reputed fossil man of the Neanderthal. *Quart. Jour. Sci.* 1:88-97.

Mayr, E. 1942. *Systematics and the Origin of Species*. Columbia University Press, New York. 334 pp.

Schwartz, J. H., & I. Tattersall. 1991. Phylogeny and nomenclature in the *Lemur*-group of Malagasy strepsirhine primates. *Anthropol. Pap. American Mus. Nat. Hist.* 69:1-18.

Simpson, G. G. 1944. *Tempo and Mode in Evolution*. Columbia University Press, New York. 237 pp.

Sturmbauer, C., & A. Meyer. 1992. Genetic divergence, speciation and morphological stasis in a lineage of African cichlid fishes. *Nature* 358:578-581.

Tattersall, I. 1986. Species recognition in human paleontology. *Jour. Hum. Evol.* 15:165-175.

_____. 1991. What was the human revolution? *Jour. Hum. Evol.* 20:77-83.

_____. 1992. Species concepts and species identification in human evolution. *Jour. Hum. Evol.* 22:341-349.

_____. 1993. Speciation and morphological differentiation in the genus *Lemur*. Pages 163-176 *in* W. H. Kimbel & L. B. Martin, eds., *Species, Species Concepts, and Primate Evolution*. Plenum Press, New York.

_____, & N. Eldredge. 1977. Fact, theory and fantasy in human evolution. *American Sci.* 65:204-211.

Thoma, A. 1970. Selektion, Gendiffusion und Spezialitätmöglichkeiten bei den Hominiden. *Homo* 21:54-60.

Wolpoff, M. H. 1973. The single species hypothesis and early hominid evolution. Pages 5-15 *in* D. Lathrap & A. Douglas, eds., *Variation in Anthropology*. Illinois Archaeological Survey, Carbondale, Illinois.

_____. 1992. *Homo erectus* in Europe: An issue of grade, of clade, or perhaps no issue at all. *Jour. Israel Prehist. Soc.*, Suppl. 1:137.

Grades and Clades:
A Paleontological Perspective
on Phylogenetic Issues

Pascal Tassy

Laboratoire de Paléontologie des Vertébrés et Paléontologie Humaine,
Université-Pierre-et-Marie Curie
4, place Jussieu
75252 Paris Cedex 05, France

The dualistic nature of fossils is responsible for the historical and current debates on the relations between paleontology, evolution, and phylogeny reconstruction. Because they ally shape (morphology) and stratigraphy (time), fossils have been considered either as evolutionary or phylogenetic facts. Though fossils, as objects, are geological facts, their interrelations — as either lineages or sister groups — are hypotheses. There is no proper paleontological approach to phylogeny. Time does not give polarity or relationships; only interpretation of morphology does. This interpretation depends on methodological choices, either phenetic or cladistic. Some aspects of debates from Lamarck to the present are historically reviewed.

Twenty years ago a paper on the relations between paleontology and phylogeny reconstruction was published (Schaeffer *et al.* 1972). This paper can be considered a landmark in the paleontological literature. Paleontologists acknowledged the heterodox view defended by some neontologists (Hennig 1950) that paleontology was not the major discipline in phylogenetic research. Briefly, the message was, "time *per se* cannot be employed in hypothesizing relationships" (Schaeffer *et al.* 1972:34). Cladistics was favored as a phylogenetic methodology whose premises were devoid of ambiguity. A consequence of cladistic inquiries was that sister-group relationships form the core of phylogenetic construction, not direct ancestor-descendant relationships.

One can argue today that little can be added, though debates have flourished since 1972. I would say roughly that, like the phoenix, the same arguments have been opposed again and again, year after year, in more or less sophisticated manners, because of key disagreements over basic concepts: What is phylogeny? What is phylogenetic information? What is the nature of paleontological information? After

Contemporary Issues in Human Evolution
Editors, W.E. Meikle, F.C. Howell, & N. G. Jablonski

Memoir 21, Copyright © 1996
California Academy of Sciences

1972, paleontologists, as a whole, did not promptly change their habits. Schaeffer *et al.* (1972:43) anticipated that "the purposeful omission of biostratigraphic information will meet with disfavor on the part of many paleontologists." Nevertheless, cladistic approaches became more frequent in the paleontological literature, especially during the last decade. Consequently, present discussions focus on the relations between cladistic analyses and the fossil record (*e.g.*, Norell & Novacek 1992; Novacek 1992b), which is undoubtedly a sign of change.

The thesis of this paper is that there is no specific paleontological approach to phylogenetic issues. On the contrary, one can observe different general approaches to phylogeny and to phylogenetic concepts, approaches that are all based on more philosophical attitudes toward the meaning of phylogeny reconstruction. These approaches influence or direct paleontological statements, that is the way we use paleontological data to infer phylogeny (Tassy 1981). Data are subordinate to the approaches. Naturally, this thesis has no implications for the question of the respective merits of data taken from extinct and extant taxa and does not imply that paleontological data are unnecessary to phylogenetic reconstruction. Fossils are crucial for a better understanding of relationships first based on extant taxa only (*e.g.*, Gauthier *et al.* 1988).

This thesis is hardly new. The relationship of fossils to the phenonenon of evolution already heated debates during the early 19th century. The various conceptions of the evolutionary process, and the triumph of gradualism later in the 19th century, heavily influenced paleontological studies in the 20th century. The distinction between pattern and process in evolutionary studies, emphasized in the cladistic literature of the 1970s, renewed not only empirical phylogenetic studies, but also old debates on the nature of phylogeny and of phylogenetic entities.

In the rapid review proposed here I will try to show, following Rudwick (1976), that the ambiguous — dualistic — nature of fossils (shape and time) is one Ariadne's clue for understanding the persistence of debates from Lamarck's day to the present.

Fossils and the Dawn of Evolutionary Thinking

I will recall briefly the two ways in which fossils were considered relative to the concept of evolution in the 19th century.

Cuvier — popularly presented as the founder of paleontology — used fossil vertebrates to demonstrate that evolution did not occur. The different past faunas described by him were interpreted as evidence of catastrophism with successive creations (Cuvier 1825). When Cuvier (1796) described his "Animal du Paraguay" (*Megatherium*), he assumed anatomical similarities between the fossil species and the extant two-toed tree sloth (*Choloepus*) and three-toed tree sloth (*Bradypus*), but — rightly — assumed no ancestor-descendant relationships between the fossil and the living taxa. Had Cuvier thought of common ancestry and not of direct ancestor-descendant relationships, the history of evolutionary research would have been quite different. But, indeed, it was then common sense to think of evolution as a synonym of direct ancestor-descendant relations.

On the contrary, Lamarck, popularly presented as the founder of the earliest rationalized theory of evolution, used fossils to demonstrate that species evolve during geologic time, so that fossil species can be linked to extant ones. There is a controversy on the role of fossils in the emergence of Lamarck's evolutionary thinking (compare Burckhardt 1977 and Laurent 1987). Although fossils play nearly no role in Lamarck's *Philosophie zoologique* (1809), fossil invertebrates studied by Lamarck himself were crucial to the emergence of the concept of transition through time from fossil species to extant ones, a decidedly evolutionary concept. In his *Mémoires sur les fossiles des environs de Paris*, published between 1802 and 1808, Lamarck described in detail the morphological variation of fossil shells in the stratigraphic context and included in the same taxonomy extant and extinct molluscs. For Cuvier & Brongniart (1811), the succession of post-Cretaceous faunas in the Paris Basin proved the interruptions of life. The study of the same period and area brought Lamarck to the opposite conclusion. One can think that it must have been easier to follow continuity through time on shells rather than on skeletons. However, if vertebrates are considered, one can recall the conclusions drawn by Geoffroy Saint-Hilaire (1831:74) from the study of fossil crocodiles from Normandie, western part of Paris Basin, and the opposite of those of Cuvier: the animals living today arose from the animals of the antediluvian world.

The different conclusions drawn by these historical figures were based on their global conception of life, probably fed, as usual, by intuition, empirical observations, and various metaphysical concerns. For example, Geoffroy Saint-Hilaire was guided by his grand view of the unity of composition, the "unité de plan," and he proposed a program for studies of fossil and recent organisms that fit with this unity: what was to be shown was how living species, by means of a kind of filiation, could be extended back to the earliest inhabitants of the earth, knowledge of which could be gained from their fragments in the fossil condition (Geoffroy Saint-Hilaire 1806:222).

Phylogeny, Grades, and Clades

Apart from the question of the role of paleontological data, the concept of phylogeny itself is an everlasting source of conflicts. The word "phylogeny" was coined by Haeckel (1866:57), originally as "the history of the development of groups (phyla)," development being understood in the modern sense of evolution. Subsequently, Haeckel (1874:18) mentioned that this "history" appeared in the dimension of geological time when he defined phylogeny as the "history of paleontological development of organic beings," as opposed to the history of individual development (ontogeny). Here "paleontological" does not imply that only fossil taxa are involved: all phylogenetic trees drawn by Haeckel include extinct and extant taxa. In the sixth edition of the *Origin of Species*, Darwin (1872:381) introduced the word "phylogeny" with the following definition: "the lines of descent of all organic beings."

I see "history" as the key word of the original definition, and I would define the concept as the history of descent of the organic beings. Yet, discussions over the decades have their origin in the double aspect of phylogeny: the pattern of branching and the amount of difference displayed by the branches. Darwin clearly contrasted the two, and, correlatively, the double nature of similarity. "The *arrangement* of the groups . . . in due subordination and relation to each other, must be strictly genealogical in order to be natural; but . . . the *amount* of difference in the several branches or groups, though allied in the same degree in blood to their progenitor, may differ greatly, being due to the different degrees of modification which they have undergone" (Darwin 1859:420). Hence we must differently rely on "strongly-marked differences in some few points, that is the amount of modification undergone" and "unimportant points, as indicating the lines of descent" (Darwin 1871:195). The well-known, single figure in the *Origin of Species* is a model: a tree in the geological dimension which exemplifies the branching aspect, that is the origin of species by splitting. Darwin contrasted "common descent" (the "arrangement" or lines of descent), and the "amount of difference" or degree of divergence. This amount does not alter the arrangement: "the amount of value of the differences between organic beings all related to each other in the same degree in blood, has come to be widely different. Nevertheless their genealogical arrangement remains strictly true, not only at the present time, but at each successive period of descent" (Darwin 1859:421).

Conflicts on the meaning of phylogeny, as well as the concepts of grade (= evolutionary levels) and clade (= monophyletic units), formalized by J. Huxley (1957), take their root in the two aspects of phylogeny detailed by Darwin. Recently, Mayr (1985:97-98) defined phylogeny as "the splitting of phyletic lineages as well as the amount by which they subsequently diverged," cladistics being the "branching of phyletic lineage," or "genealogy" (Mayr 1985:98). This semantic discussion is secondary to the question of the elucidation of relationships among organisms; nevertheless it exemplifies the weight of the two concepts.

We can read in Darwin's quotations that divergence is subordinated to the pattern of relationships. We read how the study of branching — the concept of clade — and the study of divergence — the concept of grade — will yield different evolutionary constructions with disputed relations to the concept of phylogeny.

Moreover, Darwinian and neo-Darwinian students undoubtedly favored one aspect of the evolutionary process — gradualism. Although we must bear in mind the fact that Darwin (1859:118; 1872:279, 312-313) acknowledged the possibility of periods of rapid evolution, gradualism was held as the dominant mode of evolution (Eldredge and Gould 1972, 1986). Consequently it was used as a model for most paleontological and neontological studies. Darwinism conceived evolution as adaptation through natural selection, the evolutionary process being largely gradual. One major consequence of Darwinism viewed as a research programme was the search for "laws of evolution" (T. H. Huxley 1880). Adaptation explains that the same evolutionary level (Julian Huxley's grade) can be reached independently by different groups or lineages. Hence a group, or a lineage seen as a series of ancestors and descendants, go through different evolutionary levels, or grades.

At the smallest taxonomic level a grade can be defined as a "mutation," a word introduced by the German paleontologist W. Waagen (1867) as the smallest transformation that can be reckoned in a phyletic lineage. At a higher level, we can see T.

H. Huxley's "laws of evolution" as a prefiguration of the grade concept, together with the emphasis of the phenomenon of convergence. In his famous diagram (Figure 1), the different orders of mammals are shown to reach independently different evolutionary stages, that is different grades. Lineages, grades, and evolutionary convergence, once associated in one picture, summarize the laws of evolution. Common descent then appeared to be unnecessary to explain evolution.

If paleontology has to prove evolution using fossils to show gradual change through time, we can easily conclude that the concepts of lineages and grades are adequate to fufill this program. Following Waagen, the French paleontologist Depéret used the concepts of lineage and mutation to demonstrate evolution in a strictly positivist attitude. Fossils were "facts": real organisms frozen in the rocks. A lineage — in Depéret's (1907:197) words a "rameau phylétique" — is the vertical line made from a series of fossils closely related by means of direct descent. Hence, not only were fossils facts, lineages were too. According to Depéret (1907:172), "mutations, slow and gradual" occurred in these lineages, which can be "traced back from layer to layer through the series of sedimentary stages." For Depéret a lineage is not a hypothesis or construction, and paleontology is able to identify what he calls "the real phylogenetic series." Depéret's diagram depicting the evolution of proboscideans (Figure 2) is very reminiscent of Huxley's illustration of the evolution of recent mammals. Once again, common descent does not appear to be necessary to illustrate the "fact" of evolution. The recognition in Figure 2 of the lineage "molaires à mamelons coniques" precludes any hypothesis of close relationships between tetralophodont gomphotheres (here "*M. longirostris*" and "*M. arverensis*") and elephants. As we know today, this lineage "molaires à mamelons coniques" is not a totality of descent. We can recognize without difficulty in Depéret's diagram phyletic gradualism and the basic concepts of biostratigraphy. Direct ancestor-descendant relations between fossil species was the key concept. The positivist attitude can be compared to "common sense": without ancestors, no evolution. Hence, if evolution is real, evolutionary studies are the search among fossils for real ancestors. The basis of this program was that stratigraphy enables us to see the course of evolution, or in modern words, the polarity of transformations. This is the "piège du bon sens" as Dupuis said (1986:233) or, even more roughly, "superstition" (Nelson & Platnick 1981:333). For more comments on Depéret's ideas on phylogeny see Tassy (1991).

A consequence of the emphasis on gradualism in the neo-Darwinian program is that paleontological data have been largely used as tools for illustrating phyletic gradualism or anagenesis. Stratophenetics (Gingerich 1979) reconciles two processes of evolution: cladogenesis (splitting; that is, production of taxic diversity) and anagenesis, two terms defined by Rensch (1954). Anagenesis was originally conceived for higher taxa, not only for species: the acquisition of higher levels by progressive evolution. Today, anagenesis is often understood as phyletic gradualism: transformation at the species level without production of taxic diversity. Nevertheless, stratophenetics treats fossils in the same way Depéret did: time — that is, stratigraphical position — is the key by which to interpret phenetic similarity and thus to hypothesize transformation for elucidating relationships.

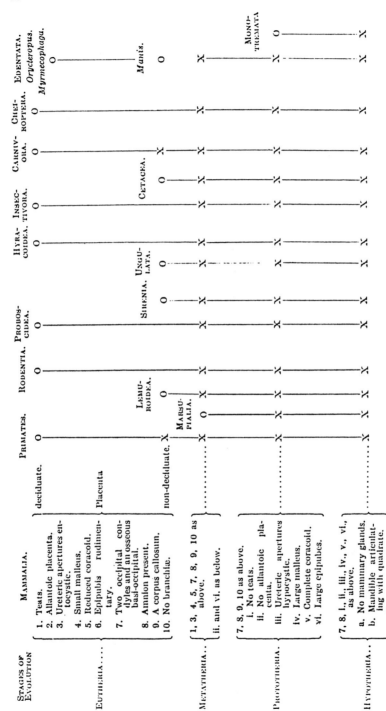

FIGURE 1. The laws of evolution applied to mammals, or how fourteen groups of mammals reached separately different grades. Taken from T. H. Huxley (1880).

	DINOTHERIUM	MASTODONTES		ÉLÉPHANTS	
		Molaires à mamelons coniques.	Molaires en crêtes transverses.	Molaires à lamelles parallèles.	Molaires à lamelles losangiques.
Époque actuelle.	»	»	»	*E. indicus.*	*E. africanus*
Époque quaternaire.	»	»	*M. americanus* . . .	*E. antiquus.*	*F. priscus.*
Époque pliocène supérieure.	»	*M. arvernensis* . . .	*M. Borsoni.*		
Époque pliocène inférieure.	»	*M. arvernensis* . . .	*M. Borsoni.*		
Époque miocène supérieure. .	{ *D. gigantissimum.* / *D. giganteum.* . .	*M. longirostris* . . .	*M. turicensis.*		
Époque miocène moyenne. . .	*D. lævius.*	*M. angustidens* . . .	*M. turicensis.*		
Époque miocène inférieure. .	*D. Cuvieri*	{ *M. angustidens, mut. pygmæus.* . . . . }	*M. turicensis.*		
Époque oligocène.	»	{ *Palæomastodon.* / *Beadnelli.*			

FIGURE 2. The five phyletic lineages of proboscideans evolving through time. Taken from C. Depéret (1907).

It can hardly be denied that the search for grades has been one of the major activities of neo-Darwinian paleontology, up to the 1980s. The evolutionary nature of the grade concept can also hardly be denied. However, it is certainly not a phylogenetic concept since it makes no reference to genealogical relationships. Huxley (1957) expressively contrasted the evolutionary concept of grade and the phylogenetic concept of clade. A clade ("branch" in Greek) is a monophyletic group, a totality of descent. For historical interest, it can be noted that, earlier, Cuénot (1940) coined the word "clade" with a somehow different meaning: an autonomous leaf of the tree corresponding to a defined structural type, that is, a *Bauplan* or fundamental body plan. I would call this a grade, notwithstanding the adjective "autonomous" which suggests, in some way, monophyly.

If clades, contrary to grades, are phylogenetic concepts, phylogenetic inquiries should focus on clades, on monophyletic groups. And here we are: the emergence of cladistics and, again, the ambiguous role of paleontology in phylogenetic studies.

Clades and Paleontology

Schaeffer *et al.* (1972) showed that paleontological data and cladistic analysis are not in conflict, although the use of the paleontological argument for polarizing characters could be misleading: chronoclines are not necessarily identical to morphoclines.

The respective roles of the different criteria for polarizing character transformations are the cause of numerous debates. The priority of outgroup comparison (Wiley 1976) or ontogenetic argument (Nelson 1973) is particularly discussed between so-called phylogenetic cladists and pattern cladists (*e.g.*, Kluge 1985; Nelson 1985). Hennig (1966) emphasized that paleontology was not per se the phylogenetic method. Nevertheless, the paleontologic argument — also called geological precedence, now acknowledged to be an auxiliary argument — was presented as the first criterion by proponents of cladistic procedures, such as Hennig (1966) himself (but not Hennig 1950) or critics such as Mayr (1986).

Hennig (1966:95) presented the paleontologic argument in the following words: if in a monophyletic group a particular character condition occurs only in older fossils, then obviously this is the plesiomorphous, and anything occuring later is the apomorphous, condition of a morphocline. This has been widely illustrated, and indeed, occurs in a vast number of cases. However, it cannot be used as a primary rule as recognized by Schaeffer *et al.* (1972). Even if it is accepted and used in priority, Hennig's definition leaves us with an unresolved question: how to identify the monophyly of the considered group? Certainly not with the criterion of geological precedence, or any morphological character seen in any Devonian taxon would be more primitive than any trait seen in any Carboniferous taxon; diversification is an old affair. This crude example explains why time does not yield relationship. The risk of circularity in the paleontological reasoning (ancient = primitive) has been pointed out many times (*e.g.*, Cracraft 1981) and acknowledged by some paleontologists (*e.g.*, Schaeffer *et al.* 1972; Goujet *et al.* 1983; Janvier 1984).

A clade is a set comprising subsets that are subclades, and the considered clade is itself a subset of a more comprehensive clade. This statement applies to clades of any rank and, if populations are accepted to be the unit of evolution, it involves species as well. As Nelson (1989:287) concludes: a species is only a taxon. The pattern of taxic diversity explains why the criterion of geological precedence is at best an auxiliary criterion. If character analysis of intrinsic characters of organisms is considered, comparative anatomy — or in more modern words: outgroup comparisons — cannot be ignored.

Figures 3-6 summarize the problems related to the paleontological criterion. Three (Fig. 3) or one (Fig. 4) morphoclines are assumed in species A-D. Lineages (Fig. 3A, 4A) are compatible with cladograms (Fig. 3B, 4B). Both represent the same distribution of characters. The number of branches strikingly differ. Lineages are "closed systems": the total information is assumed to be known; ancestors and descendants are identified. Cladograms are "open systems." They imply the existence of unknown populations that could ultimately be discovered and introduced in the cladogram without changing it. Species D could share a common ancestor with species C, situated in a place different from the basin where species A, B, C, and D, were previously found. Figure 5 shows an extreme example. Here, gaps hide the transformation series. The chronocline is the reversed morphocline. This kind of situation can be recognized if other fossils are considered which are related to the species under study and which bear some characters found in (A, B, C, D), such as a‴. This is the outgroup criterion.

The situations displayed by Figures 3-4 and Figure 5 can be mixed. In this case, chronoclines will partly fail to find the solution reached by morphocline analysis through the outgroup criterion. Figure 6 shows such a case. The transformation series a-a′ and b-b′ (Figure 6A) cannot be found by chronocline analysis. If state a′ of taxon A (chronological stage 1) is hypothesized to be ancestral for state a of taxon B, C, and D (chronological stages 2, 3, and 4), the chronocline a′-a gives the wrong solution (Figure 6B). If state b′ of taxon B (chronological stage 2) is hypothesized to be ancestral for state b of taxon C (chronological stage 3), the series b→b′→b implies homoplasy (reversal), which is the wrong solution (Figure 6B). If one admits that the fossil record is fair, and assume that the stratigraphical occurrence of taxa A, B, C, and D, reflects ancestor-descendant relationships, the taxic diversity is minimized: only one branch is recognized. In the case of Figure 6 this will be wrong, and the polarity of series a′→a, and b→b′→b will be wrong. The lineage (Figure 6B) will be less parsimonious (6 steps), while two splits imply 5 steps (Figure 6A). The criterion of geological precedence applied in priority, independantly of any other citerion, cannot identify the phylogeny displayed by Figure 6A. This example is not unrealistic, as will seen further on.

The cases illustrated by Figures 3-6 involve species. Taxonomically speaking, these species are terminal taxa that cannot be hierarchically subdivided on the basis of character analysis. If not, a terminal taxon is a set of subtaxa, and a lineage of such taxa means nothing else than a sequence of paraphyletic higher taxa, that is meta-

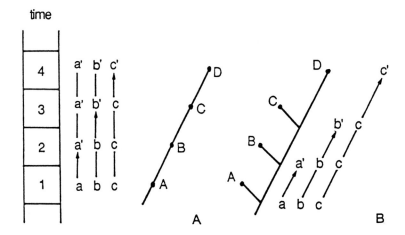

FIGURE 3. The criterion of geologic precedence. A-D: species; 1-4: geological stages; a→a', b→b', c→c': morphoclines identical to chronoclines.

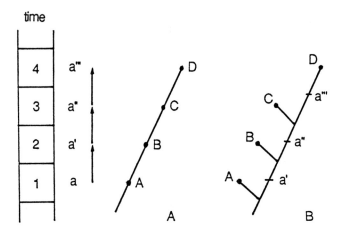

FIGURE 4. The criterion of geologic precedence. Abbreviations: see Figure 3. a→a'→a''→a''': morphocline identical to chronocline.

phorical ancestors. The criterion of geological precedence applies without ambiguity only in the case of phyletic lineages, but the prerequisite of accepting one phyletic lineage is that the fossil record is complete. If not, we are left with the possibility that diversity occurred, and accept as a competitive hypothesis a phylogenetic tree identical to the cladogram. On the basis of characters only, a phylogenetic lineage cannot be proved to be more probable than the corresponding tree. Extrinsic data are necessary, the applicability of which is conditioned by the minimization of taxic diversity: given a paleontological set, the fossil record is accepted to be complete.

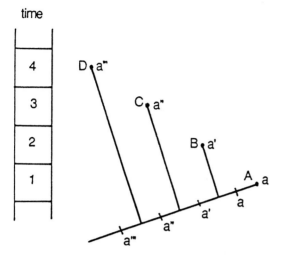

time

FIGURE 5. Refutation of the criterion of geologic precedence. Abbreviations: see Figure 3. Morphocline: a‴→a″→a′→a; Chronocline: a→a′→a″→a‴

Only in that case would ancestor-descendant relationships between two taxa (one lineage) be preferred to sister-group relationships between two taxa (two lineages).

There is a methodological necessity in minimizing evolutionary events, or steps. Otherwise any shared-derived state in various taxa could have occurred more than once, independently in each taxon. In that case common descent does not occur.

The procedure of phylogeny inference can be labelled a two-stage procedure: first, hypothesis of synapomorphy (minimization); second, construction of a pattern of taxa. I see this minimization as the inevitable sequel of the theory of descent. Certainly the procedure of minimization of evolutionary steps has consequences on the second stage (taxa and their relationships), but it does not apply directly at this stage. On the contrary, there is no methodological necessity for minimizing taxic diversity. Such a minimization would first affect the results (taxa) — that is, the second stage of the procedure — an

time

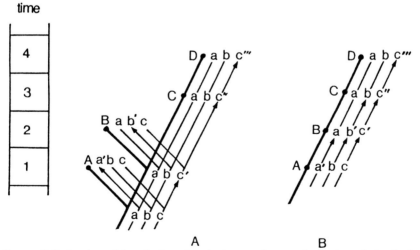

A   B

FIGURE 6. Refutation of the criterion of geologic precedence and mosaic evolution in the history of three transformation series. A: branching phylogeny; B: lineage based on chronocline analysis. Morphoclines: a→a′, b→b′, c→c′→c″→c‴. Chronoclines: a′→a, b→b′→b, c→c′→c″→c‴.

operation which is self-contradictory. Parsimony is not related to taxa (constructions) but to the evolutionary changes, that is hypotheses (here hypotheses of synapomorphy), which allow these constructions to be made.

Moreover, if character analysis is considered, a phyletic lineage implies that all characters of ancestor-descendant populations or species are known. In paleontology, this situation is the exception and not the rule, to say the least. If characters lacking due to the fragmentary nature of fossils are considered, a phyletic lineage implies that all these characters evolve without contradiction. Homoplasy, in the framework of "mosaic of characters" (de Beer 1954:48) or heterobathmy (Hennig 1966), cannot occur. As a result, this optimization goes further than the usual cladistic optimizations of unknown data. Cladistic optimizations are subordinated to character distribution on branching diagrams where homoplasy is allowed.

Minimal length trees considered in their stratigraphic framework allow for testing the adequacy of matches between cladistic events and the fossil record. Interesting comparable patterns can be provided by taxa studied at different categorical levels, with very variable time dimensions, and as different as trilobites (Edgecomb 1992), coelacanths (Cloutier 1991), and mammals (*e.g.*, Archibald 1993; MacFadden & Hulbert 1988; Novacek 1992a; Tassy 1990; Werdelin & Solounias 1991).

Figure 7 shows the pattern of differentiation of Actinistia taken from Cloutier (1991). I focus here on the sister groups labeled a and b. The extant coelacanth (*Latimeria chalumnae*) belongs to monophyletic group b: (*Hoplophagus* (*Macropoma, Latimeria*)), the earliest known member of which is late Jurassic in age. Its sister group (a) is already diversified in the early Triassic. The earliest member of this group (the genus *Alcoveria*) is separated by three dichotomies from more recent taxa. When *Alcoveria* is compared to its more closely related taxa, the chronocline appears to inversely reflect the morphocline. Taxa that branched earlier all appear later in the fossil record. This situation is comparable to that shown in Figures 5 and 6. The stratigraphical gaps are considerable, for (*Hoplophagus* (*Macropoma, Latimeria*)) more than 30% of the postulated time range (that is, 50 million years). To arrange the nine actinistian taxa so that they match with the stratigraphical sequence would alter six of the seven dichotomies.

The temporal cladogram of proboscideans shows a comparable pattern (Figure 8). This tree is based on the data matrix published by Tassy (1990) with some modifications: two characters were added to the original 136 characters, and new observations modified the mapping of seven characters. The tree displayed in Figure 8 is the strict consensus tree of three trees selected by successive weighting from 30 equally parsimonious trees (each of these 30 trees is 224 steps long, CI = 0.74, RI = 0.90). Successive weighting is an algorithm that gives *a posteriori* weight to characters according to their consistency and retention indexes. The consensus tree of the 30 equally parsimonious trees contains two multifurcations for nodes 10-13 and 15-17. These multifurcations point to uncertainties in the pattern of differentiation of Miocene Elephantoidea. In all cases (including the tree selected by successive weighting depicted by Figure 8) the postulated time-range of sister-groups shows considerable gaps. For example, sister-group relationships of *Moeritherium* imply an

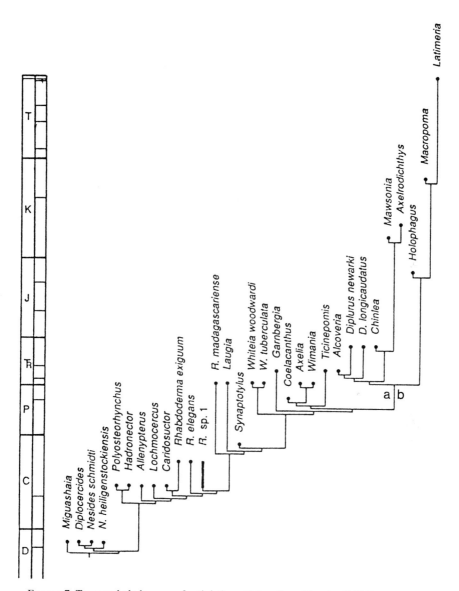

FIGURE 7. Temporal cladogram of actinistians. Taken from Cloutier (1991).

inferred range extension that covers 75% of the total range. Meanwhile, 53% and 62% of the respective ranges of *Tetralophodon* and of elephantids are inferred. The mammalian fossil record from the Miocene is often acknowledged to be fairly good. If the cladogram of Proboscidea is correct, the range of most of the tetralophodont taxa (node 14) shows a very poor record. Because of the sister-group relationship between elephantids and stegodontids (*Stegolophodon, Stegodon*) (node 18), a gap

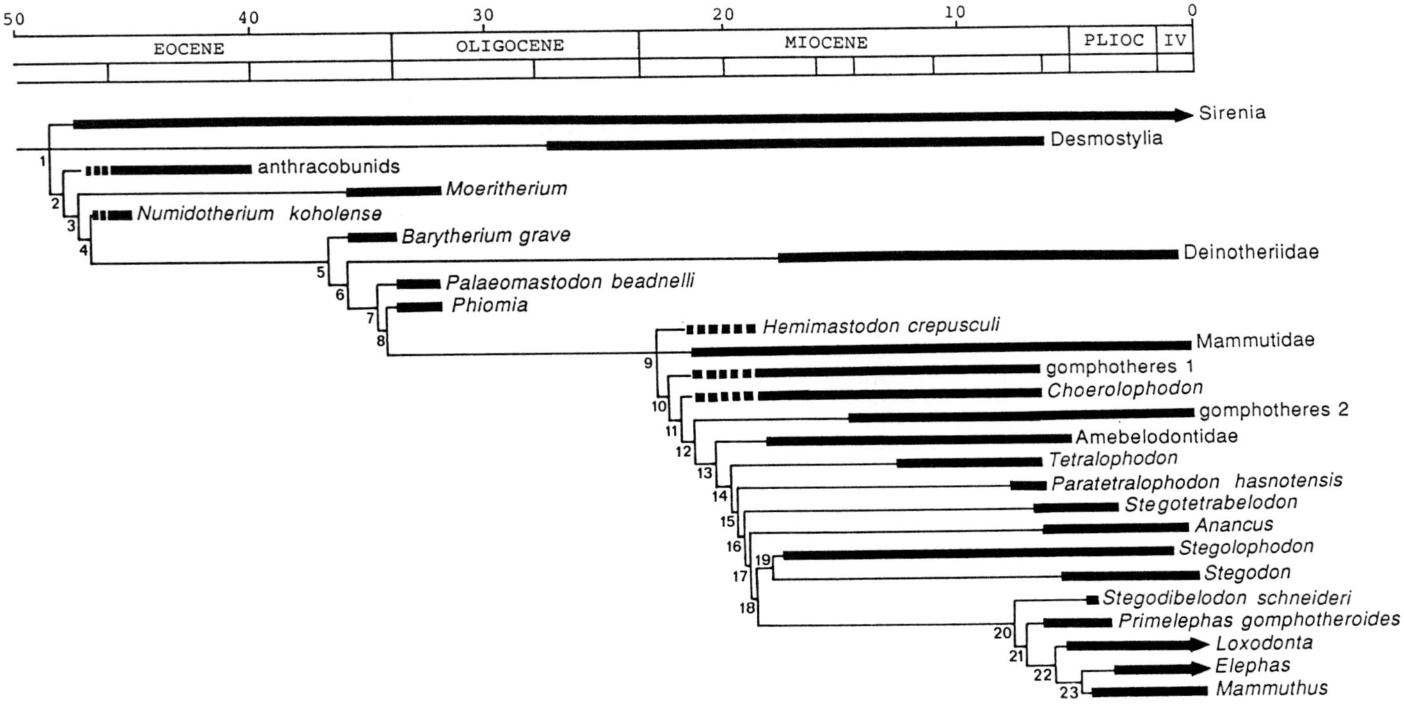

FIGURE 8. Temporal cladogram of proboscideans. Strict consensus tree of three equally parsimonious trees obtained by successive weighting (command "xs w; ie; nelsen;" of Hennig86 v. 1.5 program; length = 1447 steps, C.I. = 0.91, R.I. = 0.97).

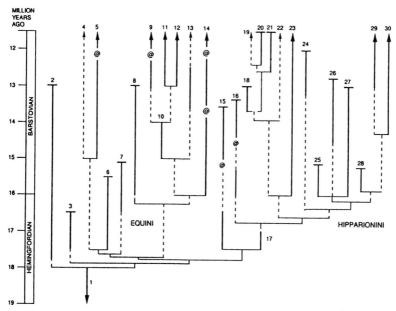

FIGURE 9. Temporal cladogram of Miocene hypsodont horses. Taken from MacFadden (1992).

of five million years between 13 Ma and 8 Ma was previously assumed (Tassy 1990). A recent discovery from the early Miocene of Thailand (Tassy *et al.* 1992) made the known range of *Stegolophodon* three to six million years older; consequently the inferred range of the elephantids is equally increased. *Stegolophodon* is a paraphyletic genus, a stem genus for *Stegodon*. Perhaps it is a stem genus for all tetralophodont forms and hypothesized nodes should be reconsidered; the gaps which affect tetralophodont taxa would be reduced. The discovery of this very early stegolophodont species yields an empirical demonstration that the fossil record is not adequate. Consequently, gaps are not sufficient to allow us to alter unambiguously a branching sequence obtained by parsimony procedures. The radiation of tetralophodont forms could have taken place in the late Early Miocene and not the late Middle Miocene as previously thought. Although we can surmise that more complete material of Early Miocene age will again change the cladogram, Figure 8 is the best fit to the data on the present evidence. (See more recent developments in Tassy 1994).

The cladistic analysis at the species level of North American equids (MacFadden & Hulbert 1988) shows again the same pattern (Figure 9). For example, sister-group relationships between the *Pliohippus mirabilis* lineage (number 5 in Figure 9) and the clade including *Equus* (the "*Astrohippus-Dinohippus* clade" of MacFadden & Hulbert, number 4 in Figure 9) implies that more than 40% of range is inferred. If one considers the details of *Astrohippus-Dinohippus* (not seen here), the morphocline contradicts the chronocline. The earliest known taxa (early Hemphillian) branched later than the later known ones (late Hemphillian) (Hulbert 1989). Here again the polarity based on stratigraphy is at odds with cladistic hypotheses of polarity.

On the other hand, during a time range of eight millions years during the Paleocene, pterychtid ungulates studied at the species level show a better fit between the cladogram and the stratigraphical record (Archibald 1993). In this example, Archibald uses Donoghue's concept of "metaspecies" (Donoghue 1985; De Queiroz & Donoghue 1988). Use of "metaspecies" (= "clade without autapomorphies" according to Archibald) leads to study of modes of speciation through the temporal cladogram of pterychtids. As a clade can only be recognized by the presence of one autapomorphy, a "clade without autapomorphies" is only meaningful if one accepts the idea that the fossil record is fairly complete, so that a paraphyletic grouping situated between two nodes of the cladogram exemplifies an ancestral species: a "metaspecies." Such an approach leads Archibald to identify modes of speciation: budding and anagenetic speciation. The cladogram of pterychtids contains 23 dichotomies; only 9 show stratigraphical gaps, and only four of them are so important that the sister-group relationships viewed in the stratigraphical context show the putative descendant older than the putative ancestor. As Archibald (1993) concludes "either the fossil record is incomplete, or the cladistic analysis is incorrect. For heuristic purposes the latter opinion is not recognized." Said differently, only the cladistic procedure allows us to identify such cases.

If we are faced by some incongruence between cladistic sequence and known stratigraphical sequence, when do we decide to change the cladistic sequence? What is the ultimate evidence that allows us to do it? One answer is stratocladistics (Fischer 1988, 1992). Stratocladistics is based on the minimization of lineages and favors ancestor-descendant relationships. A counterpart is the increase in extra steps, that is unnecessary homoplasy. Still, there is no rationale for minimizing taxic diversity. No general rule has yet been provided — and I see none — to balance the number of steps implied by cladistic analysis and the postulated gaps in the geological record, if any. These belong to separate universes.

Recent developments on this topic (*e.g.*, Huelsenbeck 1994; Salles 1995) are based on the separation of extrinsic data (time, space) from intrinsic data (shape). Cladograms are taken as guides to estimate the quality of the fossil record. In turn, different indices quantify the fit of the cladogram to the stratigraphy. Minimizing gaps in the fossil record (= minimizing "ghost lineages") can be a tool for choosing between different minimal length trees. Another way is to select non-parsimonious trees because they better fit the fossil record, even if such a procedure is based on careful statistics (*e.g.*, Wagner 1995). In that case the data (maximized homologies of the minimal length tree) are subordinated to the stratigraphic record. But no general rule is known to quantify unambiguously the quality of the record. In that case, the choice is based on vague premises such as the assumption that the record is thought to be sufficiently good to allow alterations of the cladistic result.

From these examples, can we decide when a match between stratigraphy and cladistic branching is due to chance and when it is not ? I venture to conclude that the inequality of the fossil record precludes any general rule for assessing when inadequacy is due to the possible defects of parsimony procedures.

Hominids and Concluding Remarks

The examples cited above demonstrate that stratigraphical occurrence alone cannot be used to assess polarity of characters and, consequently, relationships. Fossils have never been dismissed by cladistic analysis. On the contrary, they can be presented "as critical data for phylogeny" (Novacek 1992b). The emerging number of cladograms drawn in the stratigraphical framework but conceived only with the use of the outgroup criterion is a consequence of the nature of cladistics. Cladograms are based only on intrinsic data and are independent of extrinsic data such as known temporal range. Hence, a confrontation between the two kinds of data is possible (Norell 1992).

Phylogenetic inquiries dealing with hominids are not different from those discussed above. Early hominids were already diversified between 3 and 2 million years ago, whatever the phylogenetic hypotheses are. Three species (and perhaps more; Wood & Turner 1995) were probably contemporaneous two million years ago: two robust australopithecines and one species of the genus *Homo*. Taxic diversity occurred. The systematic pattern within the genus *Homo* itself is also probably a tree and not a single lineage. Let us consider Figure 10, a parsimony analysis of eleven characters taken in the genus *Homo* (Stringer 1987), a classic of the paleoanthropological literature. If the cladogram is viewed as a phylogenetic tree, the evolution of *Homo* displayed taxic diversity; *Homo erectus* is a grade, as is *Homo habilis*. The species *Homo sapiens* is only a clade (node 3 of Figure 10) if it includes the two geographic populations of *Homo erectus*. According to Stringer (1987) a tree closer

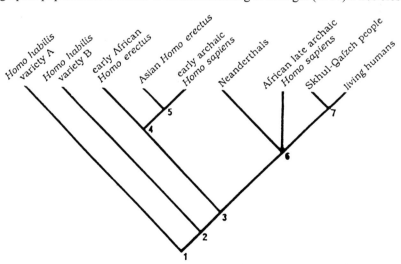

FIGURE 10. Sister-group relationships within the genus *Homo* (strict consensus tree, length = 43 steps, CI = 0.74). Taken from Stringer (1987).

to the stratigraphic sequence — in which "early archaic *sapiens*" becomes the sister group of node 6 of Figure 10 — costs three extra steps. Nevertheless, this tree could not be entirely changed into an anagenetic lineage (as in Figure 3), since diversity occurs among *sapiens*: Neandertals and fossil *Homo sapiens sapiens* ("Skhul-Qafzeh people" of node 7 of Figure 10) were contemporaneous. Nevertheless, to choose the longer tree depends on confidence in the completeness of the fossil record. To choose the shortest tree (Figure 10) depends on confidence in the matrix: this tree reflects the best fit to the data.

Cladistic procedures are said to have severe limits, and, relative to human evolution, "abundant biological limitations" (Trinkaus 1990). Still, I am not aware of any method devoid of limitations, or, better said, of prerequisites. According to Trinkaus (1990:4), cladistics can never be more than an approximation of reality because it is based on the principle of parsimony. Certainly we all wish to know what is real in our historical reconstructions. Nevertheless, science is a means to produce heuristics and representations of reality. The relation between the real world and our representation of it is an epistemologic debate into which I will not go further. Anyway, I doubt that any phylogenetic hypothesis of any sort can be more than an approximation of reality. A lineage is not a biological reality, it is a representation, a construction based on hypotheses, and so also are sister groups. Trinkaus' (1990:9) alternative to cladistics applied to human evolution is to "regard the available fossil samples (or specimens) as representative of prehistoric populations or lineages acting as portions of dynamic evolutionary units." Here we are left with the phylogenetic challenge we have discussed throughout these pages: the identification of "lineages," "portions," and "units" through reconstruction procedures.

My conclusion will be a very crude one. Paleontological data are morphological data. As such, their phylogenetic analysis cannot be distinguished from that of neontological data. Consequently, paleontological data can be analyzed in many different ways; cladistics is one of these if the aim is the reconstruction of phylogeny through character analysis. Cladistic analyses through parsimony procedures with computer programs, like PAUP (Swofford 1993) or Hennig86 (Farris 1988), have been performed for various groups of mammals, including those nearly entirely fossil. I see no reason why a mammalian group, of any rank, including Primates or even *Homo* (*e.g.*, Bonde 1977; Stringer 1987), would ontologically resist the quest for clades.

Acknowledgements

I thank Dr. L. S. Cordell, D. W. Stratmann, and the organizing committee of the Paul L. and Phyllis Wattis Foundation Endowment Symposium for inviting me to present a paper at the California Academy of Sciences. Deborah W. Stratmann's patience and efficiency were particularly appreciated. Writing the manuscript profited from Dr. D. Goujet, Dr. V. Barriel, and D. Visset's help. Figures 7, 9, and 10 were reproduced courtesy of Drs. R. Cloutier, B. MacFadden, and C. Stringer. The final manuscript was carefully edited by Dr. E. Meikle. I especially thank Prof. F.

Clark Howell for comments and suggestions on an early draft of the manuscript, and Dr. J. D. Archibald for extremely helpful criticisms. Errors remain mine.

Literature Cited

Archibald, J. D. 1993. The importance of phylogenetic analysis for the assessment of species turnover: A case history of Paleocene mammals in North America. *Paleobiology* 19:1-27.

Bonde, N. 1977. Cladistic classification as applied to vertebrates. Pages 741-804 *in* M. K. Hecht, P. C. Goody, & B. M. Hecht, eds., *Major Patterns In Vertebrate Evolution.* Plenum Press, New York.

Burkhardt, R. W. 1977. *The Spirit of System: Lamarck and Evolutionary Biology.* Harvard University Press, Cambridge, Massachusetts. 285 pp.

Cloutier, R. 1991. Patterns, trends, and rates of evolution within the Actinistia. *Environ. Biol. Fishes* 32:23-58.

Cracraft, J. 1981. Pattern and process in paleobiology: The role of cladistic analysis in systematic paleontology. *Paleobiology* 7:456-458.

Cuénot, L. 1940. Remarques sur un essai d'arbre généalogique du règne animal. *C. R. Hebd. Séances Acad. Sci.* 210:23-27.

Cuvier, G. 1796. Notice sur le squelette d'une très-grande espèce de Quadrupède inconnue jusqu' à présent, trouvé au Paraguay, et déposé au Cabinet d'Histoire naturelle de Madrid. *Magasin encycl.* 1:303-310.

———. 1825. *Discours sur les Révolutions de la Surface du Globe.* Dufour et d'Ocagne, Paris. 400 pp. (new edition: 1985, Christian Bourgois Éditeur, Paris, France. 335 pp.)

———, & A. Brongniart. 1811. Essai sur la géographie minéralogique des environs de Paris avec une carte géographique et des coupes de terrain. *Mém. Cl. Sci. Math. Phys. Inst. Imp. France* 1:1-278.

Darwin, C. 1859. *On the Origin of Species by Means of Natural Selection.* John Murray, London, UK. 513pp.

———. 1871. *The Descent of Man, and Selection in Relation to Sex.* John Murray, London, UK.

———. 1872. *On the Origin of Species by Means of Natural Selection,* 6th edition. John Murray, London, UK. 458 pp.

de Beer, G. 1954. *Archaeopteryx lithographica.* Trustees of the British Museum (Natural History), London, UK. 68 pp.

Depéret, C. 1907. *Les Transformations du Monde Animal.* Flammarion, Paris, France. 360 pp.

De Queiroz, K., & M. J. Donoghue. 1988. Phylogenetic systematics and the species problem. *Cladistics* 4:317-338.

Donoghue, M. J. 1985. A critique of the biological species concept and recommendations for a phylogenetic alternative. *Bryologist* 88:172-181.

Dupuis, C. 1986. Darwin et les taxinomies d'aujourd'hui. Pages 215-240 *in* P. Tassy, ed, *L'Ordre et la Diversité du Vivant.* Fayard/Fondation Diderot, Paris, France.

Edgecomb, G. D. 1992. Trilobite phylogeny and the Cambrian-Ordovician "event": Cladistic reappraisal. Pages 144-177 *in* M. J. Novacek & Q. D. Wheeler, eds., *Extinction and Phylogeny.* Columbia University Press, New York.

Eldredge, N., & S. J. Gould. 1972. Punctuated equilibria: An alternative to phyletic gradualism. Pages 82-115 *in* T. J. M. Schopf, ed., *Models in Paleobiology.* Freeman Cooper, San Francisco, California.

_____. 1986. Punctuated equilibrium at the third stage. *Syst. Zool.* 35:143-148.

Farris, J. S. 1988. *Hennig86 v.l.5, User's Manual.* Port Jefferson Station, New York. (Printed by the author.)

Fisher, D. 1988. Stratocladistics: Integrating stratigraphic and morphologic data in phylogenetic inference. *Geol. Soc. America, Abstracts with Program* 20(7):A186.

_____. 1992. Stratigraphic parsimony. Pages 124-129 *in* W. P. Maddison & D. R. Maddison, eds., *Excerpts from MacClade 3.* Sinauer Associates, Sunderland, Massachusetts.

Gauthier, J., A. G. Kluge, & T. Rowe. 1988. Amniote phylogeny and the importance of fossils. *Cladistics* 4:105-209.

Geoffroy Saint-Hilaire, E. 1806. Description d'un mulet provenant du canard morillon (*Anas glaucion*) et de la sarcelle de la Caroline (*Anas querquedula*). *Annls. Mus. Hist. Nat.* 7:222-227.

_____. 1831. *Recherches sur de Grands Sauriens Trouvés à l'État Fossile vers les Confins Maritimes de la Basse-Normandie.* Didot Frères, Paris, France. 138 pp.

Gingerich, P. D. 1979. The stratophenetic approach to phylogeny reconstruction in vertebrate paleontology. Pages 41-77 *in* J. Cracraft & N. Eldredge, eds., *Phylogenetic Analysis and Paleontology.* Columbia University Press, New York.

Goujet, D., P. Janvier, J. C. Rage, & P. Tassy. 1983. Structure ou modalités de l'évolution: Point de vue sur l'apport de la paléontologie. Pages 137-143 *in* J. Chaline, ed., *Modalités, Rythmes et Mécanismes de l'Évolution Biologique.* Colloq. Int. Cent. Natl. Rech. Sci. 330, Éditions du C. N. R. S., Paris, France.

Haeckel, E. 1866. *Generelle Morphologie der Organismen.* Georg Reimer, Berlin, Germany. 574 + 462 pp.

_____. 1874. *Anthropogenie oder Entwickelungsgeschichte des Menschen.* W. Engelmann, Leipzig, Germany. 732 pp.

Hennig, W. 1950. *Grundzüge einer Theorie der Phylogenetischen Systematik.* Deutscher Zentralverlag, Berlin, Germany. 370 pp.

_____. 1966. *Phylogenetic Systematics.* University of Illinois Press, Urbana, Illinois. 263 pp.

Huelsenbeck, J. P. 1994. Comparing the stratigraphic record to estimates of phylogeny. *Paleobiology* 20:470-483.

Hulbert, R. C., Jr. 1989. Phylogenetic interrelationships and evolution of North American late Neogene Equinae. Pages 176-197 *in* D. R. Prothero & R. W. Schoch, eds., *The Evolution of Perissodactyls.* Oxford University Press, New York.

Huxley, J. S. 1957. The three types of evolutionary process. *Nature* 1880:454-455.

Huxley, T. H. 1880. On the application of the laws of evolution to the arrangement of the Vertebrata and more particularly of the Mammalia. *Proc. Zool. Soc. London* 1 80:649-662.

Janvier, P. 1984. Cladistics: Theory, purpose, and evolutionary implications. Pages 39-75 *in* J. W. Pollard, ed., *Evolutionary Theory: Paths into the Future.* John Wiley & Sons, Chichester, UK.

Kluge, A. G. 1985. Ontogeny and phylogenetic systematics. *Cladistics* 1:13-27.

Lamarck, J. B. Monet de. 1809. *Philosophie Zoologique.* Dentu, Paris, France. 428 + 275 pp.

Laurent, G. 1987. *Paléontologie et Évolution en France, 1800-1859.* Èd. du Comité des travaux historiques et scientifiques, Paris, France. 553 pp.

MacFadden, B. J. 1992. Interpreting extinction from the fossil record: Methods, assumptions, and case examples using horses (family Equidae). Pages 17-45 *in* M. J. Novacek & Q. D. Wheeler, eds., *Extinctions and Phylogeny.* Columbia University Press, New York.

MacFadden, B. J., & R. C. Hulbert. 1988. Explosive speciation at the base of the adaptive radiation of Miocene grazing horses. *Nature* 336:466-468.

Mayr, E. 1985. Darwin and the definition of phylogeny. *Syst. Zool.* 34:97-98.

_____. 1986. La systématique évolutionniste et les quatre étapes du processus de classifica-
tion. Pages 145-160 *in* P. Tassy, ed., *L'Ordre et la Diversité du Vivant*. Fayard/Fondation
Diderot, Paris, France.

Nelson, G. 1973. The higher-level phylogeny of vertebrates. *Syst. Zool.* 22:87-91.

_____. 1985. Outgroups and ontogeny. *Cladistics* 1:29-45.

_____. 1989. Cladistics and evolutionary models. *Cladistics* 5:275-289.

_____, & N. Platnick. 1981. *Systematics and Biogeography: Cladistics and Vicariance*.
Columbia University Press, New York. 567 pp.

Norell, M. A. 1992. Taxic origin and temporal diversity: The effect of phylogeny. Pages
89-118 *in* M. J. Novacek & Q. D. Wheeler, eds., *Extinction and Phylogeny*. Columbia
University Press, New York.

_____, & M. J. Novacek. 1992. The fossil record and evolution: Comparing cladistic and
paleontological evidence for vertebrate history. *Science* 255:1690-1693.

Novacek, M. 1992a. Mammalian phylogeny: Shaking the tree. *Nature*, 356:121-125.

_____. 1992b. Fossils as critical data for phylogeny. Pages 46-88 *in* M. J. Novacek & Q. D.
Wheeler, eds., *Extinction and Phylogeny*. Columbia University Press, New York.

Rensch, B. 1954. *Neuere Probleme der Abstammungslehre: Die Transspezifische Evolution*.
F. Encke, Stuttgart, Germany. 407 pp.

Rudwick, M. J. S. 1976. *The Meaning of Fossils*. University of Chicago Press, Chicago,
Illinois. 287 pp.

Salles, L. de Oliveira. 1995. *Phylogénie des Ongulés Basaux: L'Évolution du Complexe
Dentaire (Morphologie des Dents Jugales Supérieures) chez les "Condylarthres" du
Crétacé supérieur à l'Eocène et l'Émergence des Cétacés (Ungulata, Mammalia); Avec la
Formulation d'un Nouvel Indice de Cohérence Strato-phylogénétique: "l'Indice de
Gaudry."* Ph.D. Dissertation, Université Paris VII, Paris, France. 477 pp.

Schaeffer, B., M. K. Hecht, & N. Eldredge. 1972. Phylogeny and paleontology. Pages 31-46
in T. Dobzhansky, M. K. Hecht, & W. C. Steere, eds., *Evolutionary Biology*, vol. 6.
Century-Crofts, New York.

Stringer, C. B. 1987. A numerical cladistic analysis for the genus *Homo*. *Jour. Hum. Evol.*
16:135-146.

Swofford, D. L. 1993. *PAUP version 3.1*. Computer program distributed by Illinois Natural
History Survey, Champaign, Illinois.

Tassy, P. 1981. Phylogeny as the history of evolution and phylogeny as a result of scientific
investigation: The significance of paleontological data. Pages 65-73 *in* J. Martinell, ed.,
International Symposium on Concept and Method in Paleontology, contributed papers.
University of Barcelona, Spain.

_____. 1990. Phylogénie et classification des Proboscidea (Mammalia): historique et actu-
alité. *Ann. Paléontol.* 76:159-224.

_____. 1991. *L'Arbre à Remonter le Temps*. Christian Bourgois Éditeur, Paris, France. 352
pp.

_____. 1994. Gaps, parsimony, and early Miocene elephantoids (Mammalia), with a reevalu-
ation of *Gomphotherium annectens* (Matsumoto, 1925). *Zool. Jour. Linnean. Soc.* 112:101-
117.

_____, P. Anupandhanant, L. Ginsburg, P. Mein, B. Ratanasthien, & V. Sutteethorn. 1992.
A new *Stegolophodon* (Proboscidea, Mammalia) from the early Miocene of northern
Thailand. *Geobios* 25:511-523.

Trinkaus, E. 1990. Cladistics and the hominid fossil record. *American Jour. Phys. Anthropol.*
83:1-11.

Waagen, W. 1867. Die Formenreihe des Ammonites subradiatus: Versuch einer paläontolo-
    gischen Monographie. *Geognost. Paläont. Beitr*. 2:181-256.
Wagner, P. J. 1995. Stratigraphic tests of cladistic hypotheses. *Paleobiology* 21:153-178.
Werdelin, L., & N. Solounias. 1991. The Hyaenidae:Taxonomy, systematics and evolution.
    Fossils Strata 30:1-104.
Wiley, E. O. 1976. The phylogeny and biogeography of fossil and recent gars (Actinopterygii;
    Lepisosteidae). *Misc. Publ. Univ. Kansas Mus. Nat. Hist*. 64:1-111.
Wood, B., & A. Turner. 1995. Out of Africa and into Asia. *Nature* 378:239-240.

Homoplasy, Clades, and Hominid Phylogeny

Henry M. McHenry
Department of Anthropology
University of California
Davis, CA 95616

Early hominid species share many unique traits, but some of these resemblances apparently evolved independently in separate lines of descent. Such parallel evolution is called homoplasy. Homoplasy tends to obscure phylogenetic relationships. Cladistic analysis helps clarify this obscuration. It also helps reveal assumptions, biases, and other problems associated with reconstructing our family tree. The resulting phylogeny shows an unexpected pattern with extensive parallel evolution in two lineages of "robust" australopithecines and a close relationship between later "robust" australopithecines and early *Homo*.

"For animals, belonging to two most distinct lines of descent," wrote Darwin (1859:427) "may readily become adapted to similar conditions, and thus assume a close external resemblance." He went on to warn "but such resemblances will not reveal – will rather tend to conceal their blood-relationship to their proper lines of descent." The fin-like limbs of whales and fishes, he observed, resemble each other not because of common descent but as adaptations for swimming. The resemblance is due to convergent or parallel evolution and not to inheritance from a common ancestor. The independent evolutionary appearance of a trait in two or more taxa, where the trait is not inherited from a common ancestor that also had that trait, is referred to as homoplasy (see Sanderson (1991) for a recent review of homoplasy). The wings of birds and bats are homoplastic. The opposable big toes of lemurs, monkeys, and apes are not because this resemblance is due to inheritance from an ancestor who also had this kind of grasping foot.

The idea is simple, but subtleties can obscure the distinction between resemblances due to homoplasy and those due to common inheritance. In the sixth edition of *On the Origin of Species* Darwin pointed out one reason for this. Closely related organisms, he observed "have inherited so much in common in their constitution, that they are apt to vary under similar exciting causes in a similar manner; and this would obviously aid in the acquirement through natural selection of parts or organs strikingly like each other, independently of their direct inheritance from a common progenitor" (Darwin 1872:427).

Among species of fossil hominids, some resemblances are due to homoplasy and some are due to descent from a common ancestor that also possessed the trait. The homoplastic resemblances are due to the fact that these species are so closely related that they evolved similarities in parallel as they adapted to the same environments. Homoplasy obscures attempts to find phylogenetic relationships. An excellent example of this obscuration is the problem of interpreting the "Black Skull" (KNM-WT 17000).

When Alan Walker found this specimen in 1985 in 2.5 million year old deposits in northern Kenya, there was little doubt that it belonged among the "robust" australopithecines.[1] It shared with *Australopithecus robustus* and *A. boisei* a suite of traits related to heavy chewing including huge cheek teeth, massive jaws, and heavily buttressed skull to withstand the chewing forces. These resemblances imply close phylogenetic affinity among these hominids, if these traits are due to descent from an ancestor that also shared them. Most authors agree that the "robust" australopithecines form a branch of our family tree that is quite separate from the lineage leading to *Homo* (see Grine (1993) for an overview of current ideas). The Black Skull and other "robust" hominids between 2.6 and 2.3 Ma form the base of this branch as the species *A. aethiopicus*. The species from South Africa, *A. robustus* (1.8-1[?] Ma), and East Africa, *A. boisei* (2.2-1.3 Ma), are the terminal parts of the "robust" branch. Relative to other contemporary hominid species, they form one branch, or, more precisely, they are monophyletic. They share a similar complex of features related to heavy chewing. Their faces, cranial vaults, jaws, and teeth are strikingly similar in many specific ways. Presumably they inherited these similarities from a common ancestor so that these resemblances are homologous.

If the "robusts" are monophyletic with *A. aethiopicus* at the base and *A. robustus* and *A. boisei* arising out of this common stem, then why do these later two species resemble early *Homo* in so many ways? Early *Homo* and the later "robust" species share numerous traits that are not present in *A. aethiopicus*. These resemblances include brain expansion, flexion of the cranial base, reduction of prognathism, deepening of the jaw joint, and a host of other features. In fact, in many ways *A. robustus* and *A. boisei* resemble early *Homo* more than any of these resembles *A. africanus* (the 3-2 Ma South African "gracile" australopithecine). Perhaps these resemblances are due to homoplasy. On the other hand, perhaps the resemblances between *A. aethiopicus* and the later "robust" australopithecines are due to homoplasy and the "robust" species are not monophyletic.

This is a difficult paradox to resolve. Some of these resemblances are concealing the true "blood-relationships to their proper lines of descent" to use Darwin's (1859:427) words. Clearly some procedure needs to be applied to parcel out homoplasy in a way that makes assumptions and biases clearly visible. Many procedures are available, but those developed by Willi Hennig (1966) have evolved into an approach that has been effectively applied to this problem (Tassy this volume). Although many extreme views have developed out of Hennig's system (Hull 1988), some fundamental features of what is commonly referred to as cladistics or phylogenetic analysis have proved to be very useful.

Cladistic Analysis

One useful feature of cladistic analysis is that when it is properly applied assumptions and biases are clearly revealed. At each step one has to expose one's thinking to critical analysis. This exposure safe-guards against the corrupting desire to advocate a fixed position.

There are many ways to proceed in a cladistic analysis, but it is helpful to follow a few basic steps. First, traits and species must be defined clearly. This step exposes a flank for critics to attack, but insures that the practitioner has some depth of defense. Paleospecies are hard to define, of course, and traits must be selected with special care. Care means that the traits are selected without bias due to preconceived notions. Care also must be given to the functional meaning of the trait. A paradox arises here because the interpreter of the functional meaning of a feature may have a bias about what the overall scheme of phylogenetic relationships "ought" to be.

A second step involves following the sequence of changes in the trait in the species under study without regard to preconceived notions about the direction of change. Brain size in human evolution increases through time, and there is no problem with bias in that. Occlusal area of cheek teeth is medium in the earliest well known species of hominid (*A. afarensis*), large in one of the next oldest species (*A. africanus*), huge in the "robust" australopithecines that came after *A. africanus*, and medium again in the earliest *Homo*. This second step disregards time and preconception. The sequence of changes in cheek tooth size places *A. afarensis* and the earliest *Homo* together.

The beauty of this formal procedure is that the practitioner is exposed at every step to corrections kindly provided by colleagues. A third step, exposing even greater vulnerability, consists of arranging the species according to where they fall from most primitive to most derived. There are formal procedures for such ordering. Time provides an imperfect clue (Tassy, this volume); usually the most primitive is the earliest, but not always. A check is provided by comparing the expression of the trait in closely related species which are not part of the analysis (outgroups, which for hominid studies consist of non-hominid members of the ape and human superfamily, Hominoidea). There are a variety of other methods of finding the direction of change for each trait, but usually time and out-group tell a consistent tale.

At this point one has a trait list with the expression of each trait in a list of species and a direction of change for each trait. From this one can derive a branching tree of relationships (cladogram) for each trait. This procedure further exposes the practitioner to scrutiny by doubting colleagues, although drawing a cladogram is quite lock-step. One simply takes each trait individually, joins the two most derived species by two intersecting lines, connects the next most derived species, and so on. The simple diagram for each trait is meaningful. The two most derived species are united by descent from a common ancestor which presumably also expressed the trait. So simple it is, but so easily missed.

The final step is to compile all the cladograms for all the traits and look for patterns. Usually there are many different cladograms. Brain size and cheek tooth area in hominid species produce two quite distinct patterns. Here again the cladist exposes the weaknesses and strengths of the analysis for all to judge. The most common way to resolve a conflict between cladograms is by choosing the one that requires the least amount of homoplasy and is the most consistent with the data (*i.e.*, the most parsimonious).

Species, Traits, and States

There are many ways to divide up the hominid bone pile into genera and species; compare, for example, Groves (1989), Tobias (1991), and Wood (1991). There is general consensus that there are three species of "robust" *Australopithecus* (*A. aethiopicus*, *A. boisei*, and *A. robustus*), but there are good reasons to accept a fourth, *A. crassidens* (Howell 1978; Grine 1982, 1985, 1988). There are also good reasons to give the "robust" australopithecines their own genus, *Paranthropus* (Grine 1988; Clarke this volume), but perhaps such an honor is ill-advised because the "robust" australopithecine species may not form a monophyletic group (Skelton and McHenry 1992; Walker et al. 1986; Leakey & Walker 1988; Walker & Leakey 1988). Reasons have also been given to divide and/or re-sort *A. afarensis* (*e.g.*, Tobias 1980; Olson 1981; Ferguson 1983; Zihlman 1985; Senut & Tardieu 1985), *A. africanus* (*e.g.*, Clarke 1988; Kimbel & White 1988), and *H. habilis* (*e.g.*, Groves 1989; Wood 1991).

Right at the start of a cladistic analysis one must lay the cards on the table and reveal choices. To make any progress, it is useful to take the attitude that one is setting up a hypothetical case in as reasonable a fashion as possible to find the logical consequences. In what follows, early hominids are divided into five species of *Australopithecus*. Early *Homo* is treated as a single species. *Ardipithecus ramidus* (White *et al.* 1994, 1995) and *Australopithecus anamensis* (Leakey *et al.* 1995) are not included because they have not yet been fully described.

Australopithecus afarensis: The earliest (3.8-2.8 Ma) well-defined species is difficult to appreciate fully because it has such a wonderful mixture of ape-like and human-like qualities. Its skull is close to what one might expect in the common ancestor of apes and people with an ape-sized brain (endocranial volume of 415 cc, which is roughly the same as a chimpanzee and not at all like modern people who average about 1350 cc), big muzzle, flat cranial base, flat jaw-joint, nasal sill, and sagittal crest that is highest in the back. Its teeth bridge the gap between ape and people with large central and small lateral upper incisors (ape-like), upper canine reduced (human-like) but still large and with shear facets formed against the lower premolar (ape-like), variable lower first premolar with some individuals having only one strong cusp (ape-like) and others with some development of a metaconid cusp (between modern ape and human), and parallel or convergent tooth rows (ape-like). Their cheek teeth were quite large relative to their body weight. Postcranially, *A. afarensis* is mostly human-like in having a hip, thigh, knee, ankle, and foot adapted to bipedality. However, superimposed on this human-like body are many traits reminiscent of the common ancestor such as somewhat elongated and curved fingers and toes, a relatively short thigh, and backwardly facing pelvic blades. Sexual dimorphism in body size is greater than in modern people, but not as high as in *Gorilla* or *Pongo*. Males weighed about 45 kg and females 29 kg. They lived in a mixed habitat, with some in well-watered and woodland conditions and others in more open environments.

Australopithecus aethiopicus: This is the least well-known of all the species, but the bits of fossils available appear to be sufficiently different to warrant separation (Grine 1988; Kimbel *et al.* 1988; see Walker *et al.* 1986 for a counter view). The species *aethiopicus* was proposed on the basis of a partial mandible from Member C of the Shungura Formation of Omo (Arambourg & Coppens 1967), but is best represented by a nearly complete cranium (KNM-WT 17000: Walker *et al.* 1986). These and isolated teeth span from 2.6 to 2.3 Ma. The cranium is much like *A. afarensis* in having an unflexed cranial base, flat jaw joint, strong prognathism, a small cranial capacity (419 cc), and other traits. These primitive traits combine with highly derived features associated with heavy chewing such as massive cheek teeth, "dished" midface with zygomatics raked forward, a smooth transition between the naso-alveolar clivus and the floor of the nose, a deep tympanic plate, and various other traits which resemble later "robust" australopithecines. It also has a heart-shaped foramen magnum like *A. boisei*.

Australopithecus africanus: Relative to *A. afarensis* and *A. aethiopicus* this species has more *Homo*-like craniodental features.[2] It is only known in South Africa, and its age is only approximately established (3-2 Ma). Its vault is higher and more rounded than the earlier species, its face is less prognathic, and its jaw joint is deeper. The lower first premolars are bicuspid. Endocranial capacity is larger (442 cc). Although the postcranium is much like *A. afarensis*, the hand bones are more *Homo*-like (Ricklan 1987, 1990). Body size resembles *A. afarensis*, although sexual dimorphism appears to be slightly reduced with males weighing about 41 kg and females 30 kg. The cheek teeth are larger than those of *A. afarensis*.

Australopithecus robustus: This species is also confined to South Africa. Its geological age is only approximately known (1.8-1[?] Ma). Its craniodental morphology is specialized for heavy chewing with very large cheek teeth, robust jaws, a flat face with cheek bones raked forward, and a sagittal crest. However, it also has many more *Homo*-like traits than earlier species: the brain is larger (530 cc), the face is not prognatic, the cranial base is strongly flexed, the jaw joint is deep, and the hands are more modern. They were relatively small-bodied (females weighed about 32 kg and males only 40 kg). They lived in a relatively dry habitat in open grasslands.

Australopithecus boisei: This species derives from East African deposits from at least 2.2 to 1.3 Ma. It is the most specialized for heavy chewing of all these species, with massive cheek teeth, jaws, and supporting architecture of the cranium to withstand the force generated by the huge chewing muscles. However, like *A. robustus* it possessed many *Homo*-like traits not seen in earlier species such as a larger brain (515 cc), flatter face, more flexed cranial base, and deeper jaw joint. Body size is poorly known, but may have been about 34 kg for females and 49 kg for males.

Homo habilis: As Wood points out (this volume), there is evidence that there may be two species represented among specimens attributed to *H. habilis*, although Tobias (1991) makes a strong case for just one. For the purposes of this study it is appropriate to regard specimens from 2.4-1.6 Ma that have been referred to as *Homo* as a single unit. Variability is high, but some consistent differences from *Australopithecus* are apparent. Brain size is larger (631 cc average), vaults are more rounded and higher, and cheek teeth are smaller. Body size for males may have been about 52 kg and for females 32 kg.

Traits and States

Table 1 (Appendix A) lists 77 variable morphological traits and their expression in each of these species and in an outgroup (extant great apes). The traits include 22 features of the face, palate, and zygomatic arch, 25 dental traits, 7 of the mandible, 10 of the basicranium, and 13 of the cranial vault. These can be grouped into five functional complexes involving heavy chewing, anterior dentition, basicranial flexion, prognathism/orthognathism, and encephalization. Skelton & McHenry (1992) provide a full description and discussion of these traits and their functional meaning. Strait *et al.* (in prep.) provide a revised and up-dated trait list including greater detail on the expressions of these and other traits in chimpanzees, gorillas, and all species of hominids except *A. ramidus* and *A. anamensis*.

Morphoclines and Cladograms

The direction of evolutionary transformation for each trait (*i.e.*, the polarity) flows from primitive to derived. The outgroup (*i.e.*, great apes) determines the primitive pole unambiguously for all of the craniodental traits in this study. For example, endocranial volume is 385 cc in *Pan* (outgroup), 415 cc in *A. afarensis*, 419 cc in *A. aethiopicus*, 442 cc in *A. africanus*, 515 cc in *A. boisei*, 530 cc in *A. robustus*, and 631 in early *Homo*. This sequence is the polarized morphocline. It implies that early *Homo* and *A. robustus* are the most highly derived for this trait and can be joined as sister taxa relative to all other species. Their joint line connects next to *A. boisei* to form a group of three species who are derived relative to all other species. The process continues until all groups are connected into a cladogram as shown in Figure 1.

The simplicity and straightforwardness of this procedure allows for conflicting evidence. For example, unlike endocranial volume, cheek tooth area goes from 294 mm^2 in the out-group, to 460 mm^2 in *A. afarensis*, to 479 mm^2 in early *Homo*, to 516 mm^2 in *A. africanus*, to 588 mm^2 in *A. robustus*, to 688 mm^2 in *A. aethiopicus*, to 799 mm^2 in *A. boisei*. This results in the cladogram displayed in Figure 2, which is incompatible with the cladogram in Figure 1. To choose the one that most likely reflects the true evolutionary relationships, one must use the principle of parsimony.

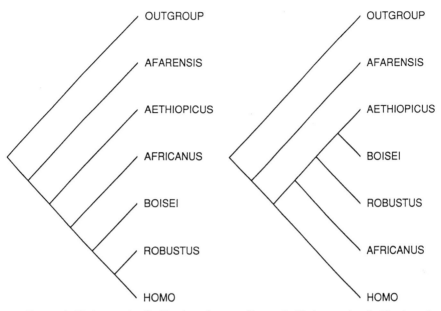

FIGURE 1. Cladogram implied by the polarized morphocline based on endocranial volume.

FIGURE 2. Cladogram implied by the polarized morphocline based on cheek tooth area.

Parsimony

The theory of parsimony is the subject of much discussion (see Sober 1988 for a review), but fundamentally it is quite straightforward. To deal with the conflicting evidence revealed by incompatible cladograms, parsimony assumes that the true phylogeny resulted from the fewest evolutionary steps. There are various measures of parsimony. The most commonly used measure is the consistency index, which is simply the minimum number of steps possible divided by the actual number of steps. If no homoplasy were present the consistency index would equal 1.

The most parsimonious cladogram that can be constructed out of the 77 traits described in Skelton & McHenry (1992) is the one shown in Figure 3. The two most derived taxa are *A. robustus* and *A. boisei*, whose stem joins *Homo* to form the next most derived group relative to the other species. The *A. robustus/A. boisei/Homo* clade then joins *A. africanus*, and that stem next joins *A. aethiopicus*. *A. afarensis* forms a sister clade to all other hominids. The consistency index is 0.722 when all 77 traits are used. This is also the most parsimonious cladogram when traits are grouped into anatomical regions or functional complexes. A cladogram linking *A. aethiopicus* to *A. boisei* and *A. robustus* as one branch and *A. africanus*/early *Homo* as another is slightly less parsimonious (CI = 0.69) using this trait list. Strait *et al.* (in prep.) show that with a revised trait list this cladogram is slightly more parsimonious than that in Figure 3.

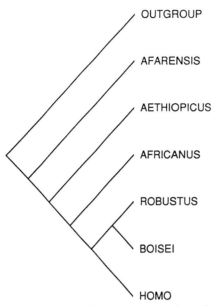

OUTGROUP

AFARENSIS

AETHIOPICUS

AFRICANUS

ROBUSTUS

BOISEI

HOMO

FIGURE 3. The most parsimonious cladogram using all 77 traits, or using summary scores from the analyses of five functional complexes or seven anatomical regions.

Our Family's Phylogeny

Figure 4 displays the phylogenetic tree implied by the most parsimonious cladogram. This phylogeny suggests that *A. afarensis* is the most primitive hominid and that all later hominids shared a common ancestor that was more derived than *A. afarensis*. This post-*afarensis* hypothetical ancestor may someday be discovered. Its morphology can be reconstructed by observing the many ways *A. aethiopicus* resembles later hominids (especially *A. africanus*) and not *A. afarensis*. For example, the upper canine jugae are prominent in the outgroup and in *A. afarensis*, but reduced or absent in all other species of hominid, which implies that the common ancestor of all post-*afarensis* species had jugae that were also reduced. This hypothetical ancestor would have a strongly developed metaconid on P_3. It would not, however resemble *A. aethiopicus* in traits related to masticatory hypertrophy (heavy chewing), nor would it resemble any other post-*afarensis* species because they are all too derived in basicranial flexion, orthognathism, and encephalization to have been the ancestor of *A. aethiopicus*.

After the divergence of *A. aethiopicus*, this phylogeny depicts a common ancestor of *A. africanus*, *A. robustus*, *A. boisei*, and *Homo* which resembled *A. africanus* in its development of anterior dentition, basicranial flexion, orthognathism, and encephalization. A second hypothetical common ancestor appears in Figure 4 to account for the numerous derived traits shared by *A. robustus*, *A. boisei*, and early *Homo* which are not seen in *A. africanus*. This ancestor would have the degree of basicranial flexion and orthognathism seen in early *Homo* and the amount of encephalization seen in *A. robustus* and *A. boisei*. This phylogeny proposes a third hypothetical ancestor which would be at the root of the lineage leading to *A. robustus* and *A. boisei*. This ancestor probably resembled *A. robustus* in traits related to heavy chewing.

These results imply that there was a large amount of parallel evolution in our family tree. The most conspicuous case of parallel evolution involves heavy chewing in *A. aethiopicus*, *A. robustus*, and *A. boisei*. This phylogeny suggests that the specific resemblances between *A. aethiopicus* and the later "robust" australopithecines are not due to descent from a common ancestor that had these traits, but due to independent acquisition. This is a very surprising result. The "Black Skull" looks so much like *A. boisei* that its discoverers and original describers attributed it to that

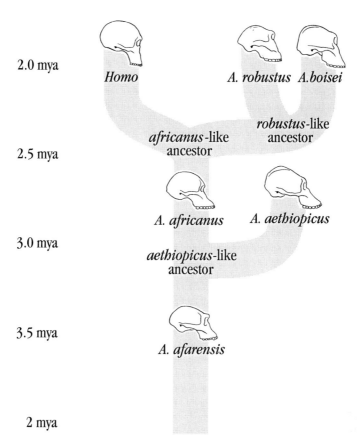

FIGURE 4. The phylogeny implied by the most parsimonious cladogram. Three hypothetical ancestors are predicted.

species and not to *A. aethiopicus* (Walker *et al.* 1986; Leakey & Walker 1988; Walker & Leakey 1988). For example, both have extreme anterior projection of the zygomatic bone, huge cheek teeth, enormous mandibular robusticity, a heart-shaped foramen magnum, and temporoparietal overlap of the occipital at asterion (at least in males).

However, all of these traits except for the heart-shaped foramen magnum are related to the functional complex of heavy chewing. The huge cheek teeth and robust mandibles of both species are obviously part of masticatory hypertrophy. The anterior projection of the zygomatic bones brings the masseter muscles into a position of maximum power. The encroachment by the root of the zygomaticoalveolar crest obscures the expression of the anterior pillars and upper canine jugae. Even the morphology of the temporoparietal overlap with the occipital is related to the function of the forces generated by the chewing muscles (see Skelton & McHenry 1992 for details).

Theoretically, it is understandable how such detailed similarity could be due to parallel evolution. This is an example of what Darwin (1872:328) referred to as species which are closely related and share "so much in common in their constitution" that similar selective forces produce similar morphologies. The selective forces in this case are related to a feeding adaptation which is associated with a specialized ecological niche. As Mayr (1969:125) points out, "Most adaptations for special niches are far less revealing taxonomically than they are conspicuous. . . Occupation of a special food niche and the correlated adaptations have a particularly low taxonomic value." In fact, many of the same traits characteristic of *A. aethiopicus* and the other "robust" australopithecines reappear in distantly related species which are adapted to heavy chewing. Expansion of the cheek teeth, shortening of the muzzle, and anterior migration of the attachment areas of the chewing muscles are seen in other primates whose diet requires heavy chewing (*e.g.*, *Hadropithecus, Theropithecus*, probably *Gigantopithecus*, and *Ekgmowechashala*).

Although the most parsimonious cladogram implies this phylogeny, other cladograms are possible but less probable. A cladogram linking *A. aethiopicus* to *A. boisei* and *A. robustus* as one branch, and *A. africanus*/early *Homo* as another, requires more evolutionary steps in this analysis because the later "robusts" resemble early *Homo* in so many features (but see Strait *et al.* (in prep.) for a different view). These include many aspects of basicranial flexion, loss of prognathism, changes in the anterior dentition, and encephalization. The postcrania, although not included in this analysis, support the view that at least *A. robustus* and early *Homo* are monophyletic relative to other species of early hominid.

Whatever the true phylogeny is, and there can be only one, the fact remains that homoplasy is commonplace. There is no avoiding it. Some resemblances appeared independently and not because of evolution from a common ancestor that possessed the same feature. Either adaptations for heavy chewing evolved twice or basicranial flexion, orthognathism, reduced anterior dentition, and encephalization evolved more than once. Darwin's astute observations apply to our own family tree.

One general lesson from this approach to hominid evolutionary biology is how to deal with ambiguity. As the King of Siam said in Rodgers and Hammerstein's *The King and I*, "What was so was so; what was not was not," but now "some things nearly so, others nearly not." It is a common experience, although perhaps uncommon to have such a clear example of ambiguity as provided by the hominid fossil record. Either heavy chewing resulted in the independent evolution of *A. aethiopicus* and *A. robustus*/*A. boisei*, or other forces shaped *A. boisei, A. robustus*, and early *Homo* to resemble each other in encephalization, basicranial flexion, anterior dentition, and orthognathism.

From this point of view it is not particularly useful to advocate a fixed position. One needs to make the best of our tiny sample of life in the past, to be open to new discoveries and ideas, and to enjoy the pleasure of learning and changing.

Acknowledgements

I thank the organizers of this symposium and particularly D. Stratmann and F. C. Howell for the invitation to contribute this paper. I am indebted to my colleague, R. R. Skelton, with whom I did the analysis that resulted in the conclusions expressed here. I am grateful to David Strait for permission to cite his manuscript which provides a valuable reanalysis of the traits, states, and species presented here. I thank all those whose work led to the discovery of the fossils, and especially the late L. S. B. Leakey, M. D. Leakey, R. E. Leakey, M. G. Leakey, F. C. Howell, D. C. Johanson, F. Thackery, C. K. Brain, P. V. Tobias, T. White, B. Asfaw, Alemu Ademasu, and Solomon Wordekal for many kindnesses and permission to study the original fossil material. I thank the curators of the comparative samples used in this study. Partial funding was provided by the Committee on Research of the University of California, Davis.

Endnotes

[1] This phrase, "robust" australopithecines, refers to early hominids that have specializations for heavy chewing. Various taxonomic names are associated with these fossils including *Paranthropus robustus* (from the South African site of Kromdraai; Broom 1938), *Paranthropus crassidens* (from Swartkrans, South Africa; Broom 1949, 1950), *Zinjanthropus boisei* (Olduvai Gorge, Tanzania; Leakey 1959), and *Paraustralopithecus aethiopicus* (Omo, Ethiopia; Arambourg & Coppens 1967). They are very similar to one another in features related to heavy chewing, and many authors prefer to recognize their similarity by designating them as a separate genus, *Paranthropus* (*e.g.*, Robinson 1972; Grine 1993; Clarke, this volume).

[2] The sample comprising *A. africanus* is heterogeneous and some specimens may belong to early *Homo* (*e.g.*, Sts 19 according to Kimbel & Rak 1993). The removal of Sts 19 from the sample does not affect the results of the analysis reported here. The *Homo*-like traits reported by Kimbel and Rak of Sts 19 do not change the scores of the relevant characters in Table 1 (*i.e.*, #57, 58, 59, 61) since there are other specimens of *A. africanus* that are *Homo*-like (*e.g.*, Sts 71, Stw 30, MLD 37/38) in their expression of one or more of these characters.

Literature Cited

Arambourg, C., & Y. Coppens. 1967. Sur la découverte, dans le Pléistocène inférieur de la vallée de l'Omo (Ethiopie), d'une mandibule d'Australopithécien. *C. R. Hebd. Séances Acad. Sci.*, Ser. D 265:589-590.

Broom, R. 1938. The Pleistocene anthropoid apes of South Africa. *Nature* 142:377-379.

——— . 1949. Another new type of fossil ape-man. *Nature* 163:57.

——— . 1950. The genera and species of the South African fossil ape-man. *American Jour. Phys. Anthropol.* 8:1-14.

Clarke, R. J. 1988. A new *Australopithecus* cranium from Sterkfontein and its bearing on the ancestry of *Paranthropus*. Pages 285-292 in F. E. Grine, ed., *Evolutionary History of the "Robust" Australopithecines*. Aldine de Gruyter, New York.

Darwin, C. 1859. *On the Origin of Species by Means of Natural Selection*. John Murray, London, UK. 490 pp.

_____ . 1872. *On the Origin of Species by Means of Natural Selection*, 6th edition. (1962 reprint) Crowell-Collier, New York.

Ferguson, W. 1983. An alternative interpretation of *Australopithecus afarensis* fossil material. *Primates* 24:397-409.

Grine, F. E. 1982. A new juvenile hominid (Mammalia: Primates) from Member 3, Kromdraai Formation, Transvaal, South Africa. *Ann. Transvaal Mus*. 33:165-239.

_____ . 1985. Australopithecine evolution: the deciduous dental evidence. Pages 153-167 *in* E. Delson, ed., *Ancestors: The Hard Evidence*. Alan R. Liss, New York.

_____ . 1988. Evolutionary history of the robust australopithecines: a summary and the historical perspective. Pages 509-570 *in* F. E. Grine, ed., *Evolutionary History of the "Robust" Australopithecines*. Aldine de Gruyter, New York.

_____ . 1993. Australopithecine taxonomy and phylogeny: historical background and recent interpretation. Pages 198-210 *in* R. L. Ciochon & J. G. Fleagle, eds., *The Human Evolution Source Book*. Prentice Hall, Englewood Cliffs, New Jersey.

Groves, C. P. 1989. *A Theory of Human and Primate Evolution*. Clarendon Press, Oxford. 375 pp.

Hennig, W. 1966. *Phylogenetic Systematics*. University of Illinois Press, Urbana, Illinois. 263 pp.

Howell, F. C. 1978. Hominidae. Pages 154-248 *in* V. J. Maglio & H. B. S. Cooke, eds., *The Evolution of African Mammals*. Harvard University Press, Cambridge, Massachusetts.

Hull, D. L. 1988. *Science as a Process*. University of Chicago Press, Chicago, Illinois. 586 pp.

Kimbel, W. H., & Y. Rak. 1993. The importance of species taxa in paleoanthropology and an argument for the phylogenetic concept of the species category. Pages 461-522 *in* W. H. Kimbel & L. B. Martin, eds., *Species, Species Concepts, and Primate Evolution*. Plenum Press, New York.

Kimbel, W. H., & T. D. White. 1988. Variation, sexual dimorphism and the taxonomy of *Australopithecus*. Pages 175-192 *in* F. E. Grine, ed., *Evolutionary History of the "Robust" Australopithecines*. Aldine de Gruyter, New York.

Kimbel, W. H., T. D. White, & D. C. Johanson. 1988. Implications of KNM-WT 17000 for the evolution of robust australopithecines. Pages 259-268 *in* F. E. Grine, ed., *Evolutionary History of the "Robust" Australopithecines*. Aldine de Gruyter, New York.

Leakey, L. S. B. 1959. A new fossil skull from Olduvai. *Nature* 184:491-493.

Leakey, M. G., C. S. Feibel, I. McDougall, & A. Walker. 1995. New four-million-year-old hominid species from Kanapoi and Allia Bay, Kenya. *Nature* 376:565-571.

Leakey, R. E. F., & A. Walker. 1988. New *Australopithecus boisei* specimens from east and west Lake Turkana. *American Jour. Phys. Anthropol*. 76:1-24.

Mayr, E. 1969. *Principles of Systematic Zoology*. McGraw-Hill, New York. 428 pp.

Olson, T. R. 1981. Basicranial morphology of the extant hominoids and Pliocene hominids: The new material from the Hadar Formation, Ethiopia, and its significance in early human evolution and taxonomy. Pages 99-128 *in* C. B. Stringer, ed., *Aspects of Human Evolution*. Taylor & Francis, London, UK.

Ricklan, D. E. 1987. Functional anatomy of the hand of *Australopithecus africanus*. *Jour. Hum. Evol*. 16:643-664.

_____ . 1990. The precision grip in *Australopithecus africanus*: Anatomical and behavioral correlates. Pages 171-183 *in* G. H. Sperber, ed., *From Apes to Angels: Essays in Honor of Phillip V. Tobias*. Wiley-Liss, New York.

Robinson, J. T. 1972. *Early Hominid Posture and Locomotion*. University of Chicago Press, Chicago, Illinois. 361 pp.

Sanderson, M. J. 1991. In search of homoplastic tendencies: statistical inference of topological patterns in homoplasy. *Evolution* 45:351-358.

Senut, B., & C. Tardieu. 1985. Functional aspects of Plio-Pleistocene hominid limb bones: implications for taxonomy and phylogeny. Pages 193-201 *in* E. Delson, ed., *Ancestors: The Hard Evidence*. Alan R. Liss, New York.

Skelton, R. R., & H. M. McHenry. 1992. Evolutionary relationships among early hominids. *Jour. Hum. Evol.* 23:309-349.

Sober, E. 1988. *Reconstructing the Past: Parsimony, Evolution, and Inference*. MIT Press, Cambridge, Massachusetts. 265 pp.

Strait, D. S., F. E. Grine, & M. A. Moniz. (in preparation). A reappraisal of early hominid phylogeny.

Tobias, P. V. 1980. *Australopithecus afarensis* and *A. africanus*: Critique and an alternative hypothesis. *Palaeontol. Afr.* 23:1-17.

_____. 1991. *Olduvai Gorge, Volume 4: The Skulls, Endocasts and Teeth of* Homo habilis. Cambridge University Press, Cambridge, UK. 921 pp.

Walker, A., & R. E. F. Leakey. 1988. The evolution of *Australopithecus boisei*. Pages 247-258 *in* F. E. Grine, ed., *Evolutionary History of the "Robust" Australopithecines*. Aldine de Gruyter, New York.

Walker, A., R. E. F. Leakey, J. M. Harris, & F. H. Brown. 1986. 2.5-MYR *Australopithecus boisei* from west of Lake Turkana, Kenya. *Nature* 322:517-522.

White, T. D., G. Suwa, & B. Asfaw. 1994. *Australopithecus ramidus*, a new species of early hominid from Aramis, Ethiopia. *Nature* 371:306-312.

_____. 1995. *Australopithecus ramidus*, a new species of early hominid from Aramis, Ethiopia — Corrigendum. *Nature* 375:88.

Wood, B. A. 1991. *Koobi Fora Research Project, Volume 4: Hominid Cranial Remains*. Clarendon Press, Oxford, UK. 466 pp.

Zihlman, A. L. 1985. *Australopithecus afarensis*: Two sexes or two species? Pages 213-220 *in* P. V. Tobias, ed., *Hominid Evolution: Past, Present and Future*. Alan R. Liss, New York.

APPENDIX A

TABLE 1. Craniodental Characteristics of Early Hominid Species[1]							
	Species[2] *and States*[3] (nd = no data)						
Characters[4]	A	B	C	D	E	F	G
FACE							
1. Nasion approaches glabella	0	0	0	1	1	1	1
2. Location of the greatest width of the nasal bones	0	1	2	0	2	2	0
3. Inferior orbital margin rounded laterally	0	0	1	0	1	0	0
4. Infraorbital foramen location	0	0	1	0	1	1	0
5. Multiple infraorbital foramina present	0	0	2	1	2	2	2
6. Anterior pillars	0	0	3	1	1	3	2
7. Upper canine jugae presence and prominence	0	0	4	1	2	4	3
8. Upper canine jugae independent of margin of nasal aperture	0	0	2	2	2	2	1
9. Projection of nasoalveolar contour beyond bicanine line	0	0	2	1	2	2	1
10. Distinct subnasal and intranasal components of nasoalveolar clivus	0	0	3	2	3	3	1
11. Anterior projection of the subnasal region relative to the nasal aperture	0	0	0	1	2	2	2
12. Nasoalveolar clivus convexity	0	0	1	2	4	4	3
ZYGOMATIC ARCH							
13. Robusticity of zygomatic bones	0	1	3	2	3	3	0
14. Relative height of anterior masseter origin by zm-alv/orb-alv index	0	1	5	4	3	2	0
15. Zygomaticoalveolar crest weakly arched in frontal view	0	0	2	2	2	1	0
16. Anterior projection of zygomatic bone	0	0	4	2	3	4	1
17. Position of anterior edge of zygomatic process origin	0	1	2	2	3	3	2
ANTERIOR DENTITION							
18. Deciduous canine shape	0	1	nd	2	2	2	2
19. I^1 incisal edge length	0	0	nd	1	2	2	1
20. Position of I^2 roots relative to margins of nasal aperture	0	0	1	1	1	1	1
21. Projection of upper canine	0	1	nd	2	3	3	3
22. Upper canine labial crown profile	0	0	nd	1	2	2	2
23. Mesial and distal contact facets on upper canine	0	0	nd	1	1	1	1
24. Mesial occlusal edge shape of lower canine	0	0	nd	1	1	1	1
25. Prominence of lingual ridge on lower canine	0	0	nd	1	1	1	2
26. Lower canine distal occlusal edge length	0	1	nd	2	3	3	3
27. Frequency of mandibular diastema	0	1	3	2	3	3	2
POSTERIOR DENTITION							
28. Postcanine tooth area (sq. mm)	0	1	5	3	4	6	2
29. Deciduous M_1 distal crown profile	0	1	nd	3	4	4	2
30. P^3 occlusal outline	0	0	3	1	3	3	2
31. Separation of P^4 cusp apices	0	1	nd	3	4	4	2
32. P_3 metaconid development	0	0	1	1	1	1	1
33. P_3 talonid height relative to protoconid	0	1	3	2	3	3	2
34. Degree of development of "robust" features of P_3	0	1	4	2	5	6	3
35. Mandibular P_3 M-D length/B-L breadth index	0	1	4	2	3	3	1

TABLE 1 (*continued*). Craniodental Characteristics of Early Hominid Species[1]							
Characters[4]	*Species*[2] *and States*[3] (nd = no data)						
	A	B	C	D	E	F	G
POSTERIOR DENTITION (continued)							
36. P_3 occlusal crown outline	0	0	1	1	1	1	1
37. P_3 occlusal wear relative to other postcanine teeth	0	0	1	1	1	1	1
38. Predominant P_3 root conformation	0	0	nd	1	1	3	2
39. Degree of development of "robust" features of P_4	0	1	4	3	5	6	2
40. Degree of mesial appression of M_1 and M_2 hypoconulids	0	0	1	3	1	3	2
41. Degree of lower molar cusp swelling	0	1	3	2	3	4	1
42. Wear disparity between buccal and lingual molar cusps	0	0	0	1	2	2	1
PALATE							
43. Flat, shallow palate	0	0	0	1	1	1	1
44. Index of palate protrusion anterior to sellion	0	0	1	3	4	5	2
45. Index of Overlap	0	1	4	2	5	6	3
46. Angle between sellion-prosthion and Frankfurt horizontal	0	0	0	1	nd	3	2
47. Index of palate protrusion to masseter	0	1	4	2	5	6	3
MANDIBLE							
48. Mandible robusticity	0	1	5	2	4	5	3
49. Orientation of mandibular symphysis	0	1	2	3	4	4	4
50. Position of mental foramen relative to mid corpus	0	1	2	3	4	3	2
51. Direction of mental foramen opening	0	1	2	1	2	2	2
52. Hollowing above and behind the mental foramen	0	1	3	2	3	3	2
53. Width of mandibular extramolar sulcus	0	0	2	1	2	2	0
54. Height of mandibular ramus origin on corpus	0	0	0	1	2	3	1
BASICRANIUM							
55. Flexion of cranial base	0	0	0	1	2	2	2
56. Distance between M_3 and the temporo-mandibular joint	0	0	0	1	2	2	2
57. Depth of mandibular fossa	0	0	0	1	2	2	2
58. Position of postglenoid relative to tympanic plate	0	0	0	1	2	2	2
59. Orientation of tympanic plate	0	1	2	3	4	4	4
60. Shape of tympanic canal	0	1	2	3	4	4	4
61. Angle of petrous bones relative to coronal plane	0	0	0	0	1	1	1
62. Heart shaped foramen magnum	0	nd	1	0	0	1	0
63. Inflection of mastoids beneath cranial base	0	0	2	1	2	2	2
64. Inclination of nuchal plane	0	1	2	2	5	4	3
CRANIAL VAULT							
65. Cranial capacity (ml)	0	1	2	3	5	4	6
66. Cerebellar lobes "tucked under" cerebrum, without lateral flare or posterior protrusion	0	nd	0	0	1	1	1
67. Branch of middle meningeal artery from which middle branch is derived	0	nd	0	0	1	1	2
68. Occipital-marginal drainage system present in high frequency	0	1	0	0	1	1	0
69. Supraorbital tori in form of "costa supraorbitalis"	0	0	2	1	2	2	0
70. Temporal lines converge to bregma	0	2	1	0	2	2	0

Characters4	A	B	C	D	E	F	G
TABLE 1 (*continued*). Craniodental Characteristics of Early Hominid Species[1]							
Species[2] and States[3] (nd = no data)							
CRANIAL VAULT (continued)							
71. Sagittal crest present, at least in males	0	0	0	1	0	0	1
72. Size of posterior relative to anterior part of temporalis	0	1	1	2	3	3	3
73. Compound temporal/nuchal crest	0	0	0	2	2	1	1
74. Asterionic notch present	0	0	0	1	1	1	1
75. Temporoparietal overlap of occipital at asterion, at least in males	0	0	1	0	0	1	0
76. Pneumatization of temporal squama	0	0	0	2	2	1	2
77. Mastoid processes inflated and projecting lateral to supramastoid crest	0	0	2	1	2	2	0

[1] This is an abridged version of Table 1 in Skelton & McHenry (1992).

[2] Species are A = Outgroup (Great Apes), B = *A. afarensis*, C = *A. aethiopicus*, D = *A. africanus*, E = *A. robustus*, F = *A. boisei*, G = early *Homo*.

[3] States are ordered multistate characterizations.

[4] Characters are grouped into anatomical regions where face is 1-12, zygomatic arch is 13-17, anterior dentition is 18-27, posterior dentition is 28-42, palate is 43-47, mandible is 48-54, basicranium is 55-64, and cranial vault is 65-77.

The Genus *Paranthropus*: What's in a Name?

Ronald J. Clarke
Palaeoanthropology Research Unit
Department of Anatomical Sciences
Medical School
University of the Witwatersrand
Johannesburg, South Africa

The hominid genus *Paranthropus* was first recognized and named by Robert Broom. Subsequent studies, especially by John T. Robinson, demonstrated that South African members of this genus possessed a highly specialized dentition, massive mandible, and related cranial architecture, all seemingly associated with a dietary adaptation which separated them from members of the *Australopithecus/Homo* lineage. Later discoveries in East and South Africa allow us to recognize at least five species of *Paranthropus: P. robustus, P. crassidens, P. boisei, P. aethiopicus*, and an as-yet unnamed species from Sterkfontein. The species of *Paranthropus* are united by their dietary specialization related to crushing and grinding hard foods, especially in dry environments. This specialization is reflected in the morphology of their cheek teeth and a cranial architecture designed to maximize the force applied through those teeth. The oldest known species of *Paranthropus*, present about 2.5 million years ago in East Africa, still retained large anterior teeth, which were greatly reduced by 2 million years ago. *Paranthropus* represents a well-adapted hominid clade which co-existed with our ancestral lineage for at least 1.5 million years.

On Tuesday, 14 June 1938, Dr. Robert Broom recovered from Kromdraai, near Sterkfontein in South Africa, the fossilized fragments of a partial skull that was to be the first find of a bizarre form of man that we now know to have existed in South Africa, Tanzania, Kenya, and Ethiopia during the period between 2.5 million and 1 million years ago. In his description of the new specimen, Broom (1938) noted that the face was flat, the incisors and canines small, the molars different in shape from those of the known small sample of Sterkfontein ape-men, and the upper P4 was larger than that of the Sterkfontein *Australopithecus*. On this basis, he "confidently" placed the skull into a new genus and species, *Paranthropus robustus*, meaning a robustly built form of man that paralleled the main line of human development. Later, Broom (1939) commented on "the remarkable degree of flattening of the lower part of the face above the incisors and canines," noting that it was mainly on this feature that he made it the type of a new genus. Subsequent and more complete discoveries of the same form of man from Swartkrans (Broom & Robinson 1952), Olduvai (Leakey 1959), Peninj (Leakey & Leakey 1964), Omo (Howell 1969), Lake Turkana (R. E. F. Leakey *et al.* 1971), and Chesowanja (Carney *et al.* 1971) not only justified Broom's confidence but also demonstrated that *Paranthropus* was an extremely bizarre and apparently highly specialized primate. Louis Leakey (1959) went so far as to place his hyper-robust cranium from Olduvai into a distinct genus, *Zinjanthropus*, and Arambourg & Coppens (1968) created the genus *Paraustralopithecus* to accommodate an early specimen from Omo in Ethiopia.

While most researchers agreed that the specimens labeled as *Paranthropus* indeed represented an unusually massive-jawed type of hominid, they did not all agree that generic distinction from *Australopithecus* was justified. At the time when only the Kromdraai specimen was known, Simpson (1945) considered *Paranthropus* to be merely a subgenus of *Australopithecus*. Even after the discovery of more complete specimens from Swartkrans in 1948, Simpson's view was followed by Oakley (1954), Howell (1955, 1968), Leakey *et al.* (1964), and Simons (1967). Doubt had been expressed by Dart (1948a and b) about the generic distinction of *Paranthropus*, which in 1951 was lumped into *Australopithecus* by Washburn & Patterson. Others who followed this classification were Le Gros Clark (1955), Dobzhansky (1962), Campbell (1963), Pilbeam & Simons (1965), Tobias (1967), Wallace (1972), Brace (1973), von Koenigswald (1973), Wolpoff (1974), Wolpoff & Lovejoy (1975), and, since then, most other authors on the subject.

Broom's view on the generic distinction of *Paranthropus* was, however, rigorously supported by the work of John Robinson (1952, 1954a and b, 1956, 1962, 1963, 1967, 1972), who not only gave sound morphological reasons for so doing but also (1962) attributed the unique cranial architecture of *Paranthropus* to its specialized dentition, which he said must be related to dietary specialization. In 1977, after detailed analysis of the craniofacial anatomy of *Paranthropus*, *Australopithecus*, and *Homo*, I found that there was no reason to dispute Broom's and Robinson's view on the generic distinction of *Paranthropus*, and I gave a list of 20 apomorphous (specialized) characters of *Paranthropus* (Clarke 1977, 1985), as follows. This definition is based on that given by Robinson (1962), with additions by the present author.

The Genus *Paranthropus*

Paranthropus is a genus of the family Hominidae characterized by the following apomorphous characters in the cranium when compared to other genera within the family:

1) A brain that is on the average larger than that of *Australopithecus*, yet not as large as that of *Homo*.
2) Formation of a lightly concave, low forehead with a frontal trigone delimited laterally by posteriorly converging temporal crests.
3) Presence of a flattened "rib" of bone across each supraorbital margin.
4) A glabella that is situated at a lower level than the supraorbital margin.

5) Formation of a central facial hollow associated with a completely flat nasal skeleton and a cheek region which is situated anterior to the plane of the piriform aperture.

6) A naso-alveolar clivus which slopes smoothly into the floor of the nasal cavity.

7) Small incisive canals which open into the horizontal surface of the nasal floor.

8) Great enlargement of premolars relative to the molars and canines.

9) Great enlargement of molars and massiveness of tooth-bearing bone.

10) Anterior teeth small when compared to premolars and molars.

11) A tendency for the maxillary canine and incisor sockets to be situated in an almost straight line across the front of the palate.

12) Formation on the naso-alveolar clivus of prominent ridges marking the central incisor sockets, but concavities marking the lateral incisor sockets.

13) Cusps of cheek teeth low and bulbous, situated closer to the center of the crown than in other hominid genera.

14) Formation of flat occlusal wear surfaces on the cheek teeth, accompanied by smoothly rounded borders between the occlusal surfaces and the sides of the crowns of the cheek teeth.

15) Virtually completely molarized lower dm1, with anterior fovea centrally situated and with complete margin.

16) Great increase in the size of the masticatory musculature and attachments, relative to the size of the skull.

17) Temporal fossa capacious and mediolaterally expanded.

18) Formation of a broad gutter on the superior surface of the posterior root of the zygoma.

19) A tendency for the palate to be shallow anteriorly and deep posteriorly.

20) Formation of either a marked pit or a groove across the zygomaticomaxillary suture of the cheek region, at least in the South African *Paranthropus*.

In view of the well-known and long-standing support for the generic separation of *Paranthropus*, it is most surprising to find that Turner & Wood (1993) claimed that Grine & Martin in 1988 and Wood in 1991 "revived arguments that the robust taxa deserve allocation to the separate genus *Paranthropus*." Turner & Wood did not mention the numerous works of John Robinson or my supportive writings, in all of which the name *Paranthropus* has been alive and well and supported by zoologically sound credentials. It is certainly welcome news that more human anatomists and physical anthropologists are coming to the belated realization that *Paranthropus* merits generic distinction, but it is to the zoologist John Robinson that credit must be given for not only recognizing this from the outset, but also for his many clear explanations of why this was so. Indeed, Robinson commented (1972:3) that his 1954 paper on the subject "is one of the very few in which the morphological evidence is evaluated at first hand from a taxonomic point of view as opposed to the many that include statements of opinion about australopithecine nomenclature in the absence of systematic taxonomic analysis of the relevant evidence."

Indeed, of all those who chose to lump *Paranthropus* into *Australopithecus*, only two (Tobias 1967; Wolpoff 1974) actually discussed Broom's and Robinson's diagnostic criteria for the genus but concluded that the morphological distinctions between *Paranthropus* and *Australopithecus* did not warrant generic separation.

The Several Species of *Paranthropus*

So what is in a name? Is it, as Mayr (1950) said, "merely a matter of taste" as to whether one recognizes a second genus, *Paranthropus*, or is it as Dobzhansky (1962) stated that "the generic category of classification is biologically arbitrary" and that it is "merely a matter of classificatory convenience how many genera one chooses to make"?

In the case of *Paranthropus*, classificatory convenience is not to be lightly dismissed, but there is a more compelling reason. The generic name which Broom gave does apply to a very distinct cluster of species of hominids belonging to one genus and distinguished from the other two genera *Australopithecus* and *Homo* by its dietary specialization. It comprised apparently several species: *Paranthropus robustus* from Kromdraai (an early South African form), *Paranthropus crassidens* from Swartkrans (a later South African form), *Paranthropus boisei* (the massive or hyper-robust East African form), and *Paranthropus aethiopicus* (the early ancestor of *P. boisei*). Whilst all of these species show the main *Paranthropus* characters or trends, they differ from each other in a few morphological features and in geographical or temporal distribution. *Homo habilis, Homo rudolfensis, Homo erectus, Homo ergaster, Homo heidelbergensis, Homo sapiens*, and *Homo neanderthalensis* also differ from each other morphologically, geographically, or temporally. Just as one can argue the pros and cons of specific separation among these *Homo* species (*e.g.*, Wood 1992; Stringer 1993), so one can argue the merits of specific separations within *Paranthropus*. One fact is, however, very obvious, and that is that *Paranthropus* is more distinct morphologically from *Australopithecus africanus* than the latter is from *Homo habilis*. It would be more justified to place *A. africanus* into the genus *Homo*, as Robinson (1972) has done, than to place *Paranthropus* into *Australopithecus*.

There would seem to be two reasons why Robinson's proposal has not been widely accepted. The first reason is that *A. africanus* as presently constituted includes, in my view, fossils of an early species of *Paranthropus* that has led many researchers to think that *A. africanus* has similarities to *Paranthropus*. I have recently claimed (Clarke 1988a and b) that the Sterkfontein assemblage actually includes two species, *Australopithecus africanus* and a larger-toothed species that has *Paranthropus*-like cranial features. It is therefore not surprising that Tobias (1967) should have considered *A. africanus* as it was then constituted to have a range of variation that incorporated some *Paranthropus* characteristics and which lent him some support for his argument that *Paranthropus* should be classified in the genus *Australopithecus*. For example, Sts 71, formerly regarded as an *Australopithecus africanus*, has the dished-face, low frontal squama, converging temporal lines, and thin supraorbital margin of a *Paranthropus*. That is, in my view, because it is a *Paranthropus*. It was specimens such as Sts 71 that had been long accepted as being *A. africanus* that led White, Johanson & Kimbel (1981) to write: "From a functional perspective, *A. africanus* crania, mandibles and teeth foreshadow the *A. robustus* + *A. boisei* character state. Specimens of *A. africanus* exhibit robust zygomatics, relatively expanded and anteriorly situated roots of the zygomatic process, tall, vertical mandibular rami, and inflated contours of the mandible corpus. They have vertical and well-buttressed posterior symphyseal regions and enlarged postcanine teeth that wear to flat occlusal platforms." They concluded that *A. africanus* was already on the *robustus* clade. With the much larger sample we now have from Sterkfontein and the growing realization that "*A. africanus*" includes two species, the uni-generic view has to be re-examined.

The second reason why *Australopithecus* has not generally been included in the genus *Homo* is that *Homo* has been considered as a big-brained, cultural hominid so far removed intellectually from the supposedly less cerebral and uncultural *Australopithecus* that the latter has to be placed in a separate genus. These assumptions are, however, not necessarily correct. We do not have a sufficiently large sample of either *A. africanus* or the earliest *Homo habilis* crania to be sure that there was no overlap in cranial capacity. Certainly in cranial morphology *Homo habilis* is not that much different from *A. africanus*. Thus Clarke (1985) wrote, "The strong similarity between *Homo habilis* crania such as O. H. 24 and StW 53 and lightly structured *A. africanus* crania such as Sts 17 and Sts 52 would seem to suggest that *H. habilis* could well have evolved from a population of lightly structured *A. africanus*." Furthermore, we cannot be sure that *A. africanus* was not culturally endowed because, by 2.5 Ma, stone tools were already in existence at sites in Ethiopia (Harris 1983; Toth & Schick 1986). We do not yet know what species of hominid was responsible for these stone tools, nor can we be sure that they were the earliest artifacts. The possibility, therefore, remains open that *A. africanus* may yet prove to have been a maker of tools.

The Dietary Adaptation

Even before this new insight concerning the presence of two species from Sterkfontein Member 4, the differences between an *A. africanus* cranium like that of Sts 5 and a *Paranthropus* cranium such as SK 48 or SK 46 were so clear-cut, so many and so profound, that it is extraordinary that those who separated generically the very similar Sts 5 (*A. africanus*) and OH 24 (*Homo habilis*) crania were not prepared to accept the generic status of *Paranthropus*. Of course they would argue that *Homo habilis* deserves generic separation because of its apparent cerebral and cultural capacities that do not seem to have been characteristic of *Australopithecus*. I would then equally argue that the highly specialized dentition, associated cranial architecture, and massive mandible of *Paranthropus* speak of a dietary specialization that separates *Paranthropus* from the *Australopithecus*/*Homo* lineage, as John Robinson deduced and vigorously maintained in numerous publications from 1954 onward.

In his paper on adaptive radiation in the australopithecines Robinson (1963) noted that as both *Paranthropus* and *Australopithecus* are hominids they have a basic skull similarity derived from a common ancestor and the fact that they are both bipedal. Beyond this, he said, the two skulls differ sharply in patterns controlled chiefly by the specializations of the dentition. He attributed the unusual architecture of the *Paranthropus* cranium to its unusual dental specializations. He observed that whilst *Australopithecus* and *Homo* have a balanced pattern of tooth size between the anterior dentition and cheek teeth, there is an imbalance in *Paranthropus* with a sudden and disproportionate size increase of cheek teeth compared to anterior dentition. In the cheek teeth of *Paranthropus*, Robinson noted "the tooth crowns are large, the enamel is thick, the occlusal surfaces large and of low relief and the root systems very well developed. The relatively great and flat occlusal surfaces and the massiveness of the postcanine teeth clearly point to a prime dietary function of crushing and grinding. The massiveness of the entire masticatory apparatus and the relatively rapid rate of wear of the teeth indicate a diet of tough material." He concluded also from missing flakes of enamel at the occlusal margins that there was grit in the food, and this suggested roots and bulbs as part of the diet. He attributed the small size of the anterior dentition to its lessened importance in the hard vegetarian diet he envisaged. By contrast, in *Australopithecus* he said there is far less emphasis on crushing and grinding, and the larger anterior dentition is consistent with an omnivorous diet. Furthermore, "the very great similarity in dentition and general skull structure between *Australopithecus* and the hominines suggests that they were basically similar in diet and behaviour" (*i.e.*, omnivores eating both vegetable food and meat). *Paranthropus*, however, did not fit the hominine pattern in dentition or skull morphology and, said Robinson, there must be an explanation for such a difference. The dietary hypothesis seemed, he said, entirely logical and, as far as he was aware, "no evidence exists which is clearly inconsistent with such an explanation."

Indeed, microwear studies of *Paranthropus* and *Australopithecus* teeth (Grine 1981, 1987) have served to confirm Robinson's hypothesis of a dietary difference between the two genera and to confirm also that *Paranthropus* was eating harder, tougher, more fibrous foods than was *Australopithecus*. Grine demonstrated that *Paranthropus* molar crowns "display significantly higher incidences of occlusal pitting, significantly shorter but wider scratches and significantly larger pits than do *Australopithecus* teeth." He also found that the scratches on the *Paranthropus* molars showed greater heterogeneity of orientation than those on *Australopithecus* crowns. From this he concluded that the diets of *Paranthropus* and *Australopithecus* were qualitatively dissimilar, that the molars of *Paranthropus* were used for more crushing and grinding activities than were those of *Australopithecus*, and that the foods processed by *Paranthropus* were substantially harder than those chewed by *Australopithecus*. Such conclusions were supported by the microwear studies on teeth of living primates by Teaford & Walker (1984) and Teaford (1985, 1986), who were able from the microwear to distinguish those primates that feed on hard objects from those that do not. By comparison of microwear features, Grine found that *Paranthropus* grouped with *Cercocebus albigena*, *Cebus apella* and *Pongo pygmaeus*, the diets of which include hard objects such as date palm seeds and kernels, palm nuts, and bark. By contrast, *Australopithecus* grouped with primates that did not eat hard objects but which concentrated on leaves.

Peters (1981) conducted research into the food plants that could have been utilized by early African hominids in both wet and dry seasons and experimented with the range of tooth pressures required to crush and utilize the food sources. He found that during times of food scarcity when very tough foods such as dry, hard berries had to be relied upon, large-toothed ape-men, particularly *Paranthropus*, were at a distinct advantage. He also found that low cusped cheek teeth like those of *Paranthropus* were better for crushing hard foods than were the higher cusped cheek teeth of *Homo*, as such cusps would tend to break. *Homo* and smaller toothed ape-men would have either had to use tools to process the food sources or more probably would have eaten different foods from those of *Paranthropus*.

Thus as Robinson said, the tooth form and accompanying cranial architecture and musculature do indicate a dietary difference between the *Australopithecus/Homo* line and *Paranthropus*. One necessary alteration to this theme is that, whereas Robinson believed *Paranthropus* existed during wet periods, more recent data indicate that *Paranthropus* lived in dry environments (Vrba 1976). This would explain the adaptation to eating hard, fibrous food such as dry berries, but as Peters (1981) observes it does not mean that *Paranthropus* did not eat other foods as well. It rather means that *Paranthropus* could utilize foods not accessible to other hominids during dry conditions.

The most detailed analysis of a *Paranthropus* cranium and dentition was that of Tobias (1967) on the well-preserved and almost complete cranium of "*Zinjan-thropus*," Olduvai Hominid 5. Hence his conclusions on the taxonomy of the australopithecines have had considerable influence on the opinions of his colleagues concerning the status of *Paranthropus*. With the 29 years of research and discovery that have passed since publication of his monograph, it behooves us to look again at Tobias's arguments. Although Tobias (1967) argued against the dietary hypothesis and against generic distinction of *Paranthropus*, a reading of his arguments shows that he agreed that "there is a real difference between the two taxa in the disparity between the sizes of the canines and the cheek teeth," and he agreed that *Paranthropus* has larger cheek teeth than *Australopithecus*. Furthermore, he did not disagree that there are differences in cranial morphology between the two taxa. However, he attributed six of the cranial distinctions to the enlarged teeth and related structures and muscles of *Paranthropus*.

Whilst it is true that enlarged cheek teeth are a characteristic trend of *Paranthropus*, these cannot be considered as the main contributing factor to the unusual cranial architecture, as we know that even relatively small toothed females of *Paranthropus* such as SK 48 or KNM-ER 732 (R. E. F. Leakey *et al.* 1972) have the same cranial architecture. This point was discussed at length by Clarke (1977) when he criticized the use of the colloquial terms "robust australopithecine" for *Paranthropus* and "gracile australopithecine" for *A. africanus*. He concluded, "In view of the confusion which can be caused by the use of the terms 'gracile australopithecine' and 'robust australopithecine,' the present writer does not use such terms in this thesis and would advocate that they be dropped from the anthropological vocabulary."

The unusual cranial architecture of *Paranthropus* is rather due, as Robinson observed, to the unusual dental specialization which is manifested by the relatively enlarged premolars (with molarized lower dm1) and the thick-enameled, low-cusped flat occlusal surfaces of the cheek teeth. These latter three characters which are crucial elements in the dietary hypothesis did not feature in Tobias's argument. Nevertheless, he concluded that his arguments demonstrated that "With the attenuation and, indeed, collapse of the dietary hypothesis, it would seem that the main prop for the generic distinctness of the two taxa falls away too" (Tobias 1967:228). On the contrary, it would seem that the aforementioned case for generic distinction was actually strengthened by the discovery, and Tobias's analysis, of Olduvai Hominid 5, by subsequent discoveries in Tanzania, Ethiopia, and Kenya, and by the research on microwear by Grine and on food resources by Peters.

The Ancestry of *Paranthropus*

Both *Paranthropus crassidens* of Swartkrans and *Paranthropus boisei* of East Africa seem to have existed between 1.8 and 1 million years ago; *Paranthropus robustus* of Kromdraai may be a little older than this (Vrba 1982; Vrba & Panagos 1982). It had been generally assumed, especially by those who classify them as *Australopithecus,* that they evolved out of *A. africanus* which dates to around 2.5 to 3.0 million years ago from Sterkfontein, Taung, and Makapansgat. Robinson (1963, 1972), however, did not accept this concept but believed conversely that *A. africanus* evolved out of an early form of *Paranthropus* that had large anterior teeth. In this belief he was in one respect remarkably prophetic because there are now newly discovered fossils which indicate that an ancestral *Paranthropus* with large anterior teeth existed in East Africa 2.5 million years ago. However, his belief that *A. africanus* evolved out of such an ancestor does not seem to fit with the fossil data we now have in the form of 3 to 4 million year old *Australopithecus afarensis* that, with its relatively small cheek teeth, seems a more plausible ancestor to *A. africanus.*

From West Lake Turkana in Kenya has been discovered a 2.5 million year old cranium of a hominid (WT 17000) that has all the hallmarks of *Paranthropus boisei*, but with a more prognathic muzzle and larger canines and incisors (Leakey & Walker 1988). From Sterkfontein, Clarke (1988a and b) has recognized the existence of a second species of hominid alongside *A. africanus* at a date of between 2.5 and 3.0 million years ago. This second species, exemplified by StW 252, also has *Paranthropus* features in its thin supraorbital margin, sagittally converging temporal lines enclosing a slight supraglabellar hollow, broad interorbital distance, flat face, large molars and premolars. Yet, unlike *Paranthropus*, it has large canines and incisors indicating that it still retains these ancestral pongid-like features whilst specializing in its enlarged cheek teeth. This would be expected in an early form of *Paranthropus* because it is temporally nearer to a common hominid ancestor. First would come the dietary specialization with enlargement of cheek teeth. Only when the large, projecting canines and incisors became both unnecessary and a hindrance to side-to-side grinding would they become reduced in size and prominence.

Conclusions

Robinson (1972) devoted considerable discussion to the concept of the genus, stating that "a genus represents a clearly defined adaptive zone or way of life, within which various species can occur that represent no more than variations in detail on the basic adaptive theme." In support of this concept, he quoted a definition by taxonomists Mayr, Linsley & Usinger (1953) which is worth repeating here: "The genus, as seen by the evolutionist, is a group of species that has descended from a common ancestor. It is a phylogenetic unit. . . . The genus, however, has a deeper significance. Upon closer examination, it is usually found that all the species of a genus occupy a more or less well-defined ecological niche. The genus is thus a group of species adapted for a particular way of life. . . . On this theoretical basis, it is probable that all generic characters are either adaptive or correlated with adaptive characters."

It is clear that the fossils assigned to the genus *Paranthropus* do fit the above definition in that they belong to a cluster of hominid species that were united in their dietary specialization, in which they had evolved relatively large, low cusped and smooth margined cheek teeth designed to crush and grind hard foods, especially in dry environments. The unusual cranial architecture evolved at the same time to maximize the force applied through the cheek teeth in both vertical (crushing) and lateral (grinding) movements. The earliest known species of *Paranthropus*, found in East Africa, lived 2.5 million years ago and still retained large incisors and canines that were to become greatly reduced by about 2 million years ago.

There were different species of *Paranthropus* living in East and South Africa at the same time as the smaller-toothed *Australopithecus/Homo* lineage. There was thus a similar situation to that within the Carnivora, where both the East African and South African species of the genus *Hyaena* occupy the same areas as the genus *Crocuta*. The cranial and dental anatomy of *Hyaena* and *Crocuta* resemble each other to a greater degree than do those of *Paranthropus* and *Australopithecus*, and yet there is no doubt that *Hyaena* and *Crocuta* belong to distinct genera.

Clarke (1977) noted a parallel situation to that of *Australopithecus* and *Paranthropus* which can be seen among the primates of the grade Strepsirhini: "There are great similarities between the skulls of *Galago* and *Perodicticus* or *Galago* and *Arctocebus* (Hill 1953) but they have totally different skeletal anatomy. They all belong to the Family Lorisidae, but *Perodicticus* and *Arctocebus* are slow climbers, of the Subfamily Lorisinae whilst *Galago* is a vertical clinger and leaper of the Subfamily Galaginae (Napier & Napier 1967). Although *Perodicticus* and *Arctocebus* have similar skulls, skeleton and locomotor pattern, there are slight differences in their anatomy and behavior which have been considered sufficient to separate them at the generic level."

Thus the name *Paranthropus*, meaning "parallel to man," refers to a very important, well-adapted and unusual primate that existed for a period of at least one and a half million years alongside our own ancestral lineage.

Literature Cited

Arambourg, C., & Y. Coppens. 1968. Découverte d'un australopithécien nouveau dans les gisements de l'Omo (Éthiopie). *South African Jour. Sci.* 64:58-59.

Brace, C. L. 1973. Sexual dimorphism in human evolution. *Yearb. Phys. Anthropol.* 16:31-49.

Broom, R. 1938. The Pleistocene anthropoid apes of South Africa. *Nature* 142:377-379.

_____. 1939. A restoration of the Kromdraai skull. *Ann. Transvaal Mus.* 19:327-329.

_____ , & J. T. Robinson. 1952. Swartkrans ape-man, *Paranthropus crassidens*. *Transvaal Mus. Mem.* no. 6..

Campbell, B. G. 1963. Quantitative taxonomy and human evolution. Pages 50-74 *in* S. L. Washburn, ed., *Classification and Human Evolution*. Aldine, Chicago, Illinois.

Carney, J., A. Hill, J. A. Miller, & A. Walker. 1971. Late australopithecine from Baringo District, Kenya. *Nature* 230:509-514.

Clark, W. E. Le Gros. 1955. *The Fossil Evidence for Human Evolution*. University of Chicago Press, Chicago, Illinois. 181 pp.

Clarke, R. J. 1977. *The Cranium of the Swartkrans Hominid, SK 847 and Its Relevance to Human Origins*. Ph.D. Thesis, University of the Witwatersrand, Johannesburg, South Africa. 325 pp.

_____ . 1985. *Australopithecus* and early *Homo* in southern Africa. Pages 171-177 *in* E. Delson, ed., *Ancestors: The Hard Evidence*. Alan R. Liss, New York.

_____ . 1988a. Habiline handaxes and paranthropine pedigree at Sterkfontein. *World Archaeol.* 20:1-12.

_____ . 1988b. A new *Australopithecus* cranium from Sterkfontein and its bearing on the ancestry of *Paranthropus*. Pages 285-292 *in* F. E. Grine, ed., *Evolutionary History of the "Robust" Australopithecines*. Aldine de Gruyter, New York.

Dart, R. A. 1948a. The Makapansgat proto-human *Australopithecus prometheus*. *American Jour. Phys. Anthropol.* 6:259-284.

_____ . 1948b. The adolescent mandible of *Australopithecus prometheus*. *American Jour. Phys. Anthropol.* 6: 391-412.

Dobzhansky, T. 1962. *Mankind Evolving*. Yale University Press, New Haven, Connecticut. 381 pp.

Grine, F. E. 1981. Trophic differences between "gracile" and "robust" australopithecines: A scanning electron microscope analysis of occlusal events. *South African Jour. Sci.* 77:203-230.

_____ . 1987. The diet of South African australopithecines based on a study of dental microwear. *L'Anthropologie* 91:467-482.

Harris, J. W. K. 1983. Cultural beginnings: Plio-Pleistocene archaeological occurrences from the Afar, Ethiopia. *African Archaeol. Rev.* 1:3-31.

Hill, W. C. O. 1953. *Primates vol. I: Strepsirhini*. Edinburgh University Press, Edinburgh, UK. 798 pp.

Howell, F. C. 1955. The age of the australopithecines of southern Africa. *American Jour. Phys. Anthropol.* 13:635-662.

_____ . 1968. Omo research expedition. *Nature* 219:567-572.

_____ . 1969. Remains of Hominidae from Pliocene/Pleistocene formations in the lower Omo basin, Ethiopia. *Nature* 223:1234-1239.

Leakey, L. S. B. 1959. A new fossil skull from Olduvai. *Nature* 184:491-493.

_____ , & M. D. Leakey. 1964. Recent discoveries of fossil hominids in Tanganyika: at Olduvai and near Lake Natron. *Nature* 202:5-7.

_____ , P. V. Tobias, & J. R. Napier. 1964. A new species of the genus *Homo* from Olduvai Gorge. *Nature* 202:7-9.

Leakey, R. E. F., J. M. Mungai, & A. C. Walker. 1971. New australopithecines from East Rudolf, Kenya. *American Jour. Phys. Anthropol.* 35:175-186.

_____ , J. M. Mungai, & A. C. Walker. 1972. New australopithecines from East Rudolf, Kenya (II). *American Jour. Phys. Anthropol.* 36:235-251.

_____ , & A. Walker. 1988. New *Australopithecus boisei* specimens from east and west Lake Turkana, Kenya. *American Jour. Phys. Anthropol.* 76:1-24.

Mayr, E. 1950. Taxonomic categories in fossil hominids. *Cold Spring Harbor Symp. Quant. Biol.* 15:109-118.

_____, E. G. Linsley, & R. L. Usinger. 1953. *Methods and Principles of Systematic Zoology.* McGraw-Hill, New York. 336 pp.

Napier, J. R., & P. H. Napier. 1967. *A Handbook of Living Primates.* Academic Press, London, UK. 456 pp.

Oakley, K. P. 1954. Dating of the Australopithecinae of Africa. *American Jour. Phys. Anthropol.* 12:9-23.

Peters, C. R. 1981. Robust vs. gracile early hominid masticatory capabilities: The advantages of the megadonts. Pages 161-181 *in* L. L. Mai, E. Shanklin, & R. W. Sussman, eds., *The Perception of Evolution: Essays Honoring Joseph B. Birdsell.* Anthropology U.C.L.A., Los Angeles, California.

Pilbeam, D. R., & E. L. Simons. 1965. Some problems of hominid classification. *American Sci.* 53:237-259.

Robinson, J. T. 1952. The australopithecines and their evolutionary significance. *Proc. Linnean Soc. London* 163:196-200.

_____ . 1954a. The genera and species of the Australopithecinae. *American Jour. Phys. Anthropol.* 12:181-200.

_____ . 1954b. Phyletic lines in the prehominids. *Zeits. Morphol. Anthropol.* 46: 269-273.

_____ . 1956. *The Dentition of the Australopithecinae.* Transvaal Mus. Mem. no. 9.

_____ . 1962. The origin and adaptive radiation of the australopithecines. Pages 120-140 *in* G. Kurth, ed., *Evolution und Hominisation.* Gustav Fischer Verlag, Stuttgart, Germany.

_____ . 1963. Adaptive radiation in the australopithecines and the origin of man. Pages 385-416 *in* F. C. Howell & F. Bourlière, eds., *African Ecology and Human Evolution.* Aldine, Chicago, Illinois.

_____ . 1967. Variation and the taxonomy of the early hominids. Pages 69-100 *in* T. Dobzhansky, M. K. Hecht, & W. C. Steere, eds., *Evolutionary Biology.* Appleton-Century-Crofts, New York.

_____ . 1972. *Early Hominid Posture and Locomotion.* University of Chicago Press, Chicago, Illinois. 361 pp.

Simons, E. L. 1967. The significance of primate paleontology for anthropological studies. *American Jour. Phys. Anthropol.* 27:307-332.

Simpson, G. G. 1945. The principles of classification and a classification of mammals. *Bull. American Mus. Nat. Hist.* 85:1-350.

Stringer, C. 1993. Secrets of the pit of the bones. *Nature* 362:501-502.

Teaford, M. F. 1985. Molar microwear and diet in the genus *Cebus. American Jour. Phys. Anthropol.* 66:363-370.

_____ . 1986. Dental microwear and diet in two species of *Colobus.* Pages 63-66 *in* J. G. Else & P.C. Lee, eds., *Primate Ecology and Conservation.* Cambridge University Press, Cambridge, UK.

_____ , & A. C. Walker. 1984. Quantitative differences in dental microwear between primate species with different diets and a comment on the presumed diet of *Sivapithecus. American Jour. Phys. Anthropol.* 64:191-200.

Tobias, P. V. 1967. *Olduvai Gorge vol. 2: The Cranium and Maxillary Dentition of* Australopithecus (Zinjanthropus) boisei. Cambridge University Press, Cambridge, UK. 264 pp.

Toth, N., & K. D. Schick. 1986. The first million years: the archaeology of protohuman culture. Pages 1-96 *in* M. B. Schiffer, ed., *Advances in Archaeological Method and Theory,* vol. 9. Academic Press, Orlando, Florida.

Turner, A., & B. Wood. 1993. Taxonomic and geographic diversity in robust australopithecines and other African Plio-Pleistocene larger mammals. *Jour. Hum. Evol.* 24:147-168.

Von Koenigswald, G. H. R. 1973. *Australopithecus, Meganthropus*, and *Ramapithecus. Jour. Hum. Evol.* 2:487-491.

Vrba, E. S. 1976. *The Fossil Bovidae of Sterkfontein. Swartkrans and Kromdraai.* Transvaal Mus. Mem. no. 21.

_____ . 1982. Biostratigraphy and chronology, based particularly on Bovidae of southern African hominid associated assemblages: Makapansgat, Sterkfontein, Taung, Kromdraai, Swartkrans; also Elandsfontein (Saldanha), Broken Hill (now Kabwe) and Cave of Hearths. Pages 707-752 in *Prétirage, 1er Congrès Internat. Paléontol. Humaine.* CNRS, Nice, France.

_____ , & D. C. Panagos. 1982. New perspectives on taphonomy, palaeoecology and chronology of the Kromdraai apeman. Pages 13-26 *in* J. A. Coetzee & E. M. van Zinderen Bakker, eds., *Palaeoecology of Africa and the Surrounding Islands*, vol. 15. A. A. Balkema, Rotterdam.

Wallace, J. A. 1972. *The Dentition of the South African Early Hominids: A Study of Form and Function.* Ph.D. Thesis, University of the Witwatersrand, Johannesburg, South Africa. 244 pp.

Washburn, S. L., & B. Patterson. 1951. Evolutionary importance of the South African "man-apes." *Nature* 167:650-651.

White, T. D., D. C. Johanson, & W. H. Kimbel. 1981. *Australopithecus africanus*: Its phyletic position reconsidered. *South African Jour. Sci.* 77:445-470.

Wolpoff, M. H. 1974. The evidence for two australopithecine lineages in South Africa. *Yearb. Phys. Anthropol.* 17:113-139.

_____ , & C. O. Lovejoy. 1975. A re-diagnosis of the genus *Australopithecus. Jour. Hum. Evol.* 4:275-276.

Wood, B. 1992. Origin and evolution of the genus *Homo. Nature* 355:783-790.

Origin and Evolution of the Genus *Homo*

Bernard Wood

Hominid Palaeontology Research Unit
Department of Human Anatomy and Cell Biology
The University of Liverpool
PO Box 147
Liverpool L69 3BX , UK

A subset of hominid fossils recovered from East African sites, and dating to 1.5 Ma or earlier, has been assigned to the genus *Homo*. Three species are identified within this material, herein called "early *Homo*." One species resembles *Homo erectus* and is either interpreted as "early African" *H. erectus*, or *Homo ergaster*. Its incorporation within the genus *Homo* is well justified and widely supported by evidence from both phenetic and cladistic studies. The two other species, *Homo rudolfensis* and *Homo habilis*, are taxonomically more ambiguous. *Homo rudolfensis* has an absolute and relative brain size comparable to *H. ergaster*, but is only indirectly associated with postcranial remains that are evidently from a biped. Its masticatory anatomy is reminiscent of the australopithecines. The cranial base of *Homo habilis* is relatively foreshortened and wider than that of australopithecine crania, trends that are continued in later *Homo*. However, its absolute brain size is modest and its postcranial skeleton is australopithecine-like. If obligatory bipedalism is a *Homo* prerequisite, then *H. habilis* should not be included in that genus. The relationship between *A. africanus* and *Homo* is not clear. Cladistic analyses based on craniodental characters narrowly favor it being the sister group of *Homo* and thus strengthen claims for its inclusion in that genus. However, there is an equally compelling case to regard early African *Homo erectus* as the ancestral species of *Homo*, and thus to restrict the definition of the genus. The period between 2.5 and 1.5 Ma saw the emergence of an unambiguous *Homo* lineage. However, the combinations of characters within *H. rudolfensis* and *H. habilis* suggest that it was possible to acquire some of the characteristics of *Homo* and not others. We have yet to understand the context within which these various evolutionary initiatives were played out.

There is a widely-held but probably unrealistic expectation that the boundaries of the genus *Homo* will prove to be well-defined. This expectation ignores the reality which is that genera are "necessarily more arbitrary" than species (Simpson 1961). This proposition is borne out by most of the diagnoses of *Homo* which have been offered from time to time. These usually contain a summary of the features of the species which the authors judge should be subsumed within *Homo* as well as references to morphological trends such as an enlarged brain, reduced dentition, etc., which the species are judged to share.

For many years the only fossil evidence which helped to bridge the substantial divide between the morphology of modern humans, on the one hand, and that of the living African apes, on the other, was attributed to *Homo erectus*. Dentally and postcranially, its affinities with living *Homo sapiens* were evident, despite its low and heavily buttressed cranium with an average absolute brain size substantially smaller than that of modern human populations.

The discovery, in 1924 and thereafter, of fossils which are now usually assigned to *Australopithecus*, or to *Australopithecus* and *Paranthropus* (see Wood (1991) and Grine (1993) for up-to-date reviews of this material) provided evidence of creatures which were more ape-like than *H. erectus*. Compared to living apes, australopithecines demonstrate only a modest degree of brain enlargement, and while they evidently placed more emphasis on bipedal posture and locomotion than living apes, their postcranial skeletons still retain many features linked with climbing. Their anterior dentition was reduced and their muzzles were foreshortened compared to the living apes, but the postcanine teeth of the australopithecines are large relative to those of the living apes. Contemporary commentators were unwilling to amend their diagnosis of *Homo* to allow for the inclusion of the new fossils within that genus (but see Mayr (1950) and Robinson (1972) for contrary views) and instead chose the genus names *Australopithecus* (meaning "southern ape") and *Paranthropus* (meaning "beside man") which served to emphasize the ape-like features of the new material.

In the last three decades, fossils have been discovered which helped to close the remaining morphological gap between the australopithecines and *H. erectus*. These remains, attributed formally or informally to *H. habilis*, are a morphological palimpsest. Whereas their brains and teeth were intermediate between the australopithecines and *H. erectus*, postcranially they were judged to be decidedly *Homo*-like.

The taxonomic options for this new material were either to relax the diagnosis of *Homo* to allow the inclusion of a species-group with more ape-like features than *H. erectus*, or to broaden the diagnosis of *Australopithecus* or *Paranthropus* to allow the inclusion of creatures with more *Homo*-like features such as obligatory bipedalism and a reduced dentition.

This paper briefly reviews the fossil evidence for what is now often called "early *Homo*" and sets out options for its taxonomic interpretation. It then explores the relationships between the component species, examines the grounds for including each of them within *Homo*, summarizes the case for *Homo* monophyly, and then surveys the hominid fossil record for likely ancestors of the genus. Finally, the various early *Homo* taxa are scrutinized for evidence about their evolutionary grade. Do they all belong to the same locomotor and dietary grade, or is there evidence of any grade shift within the material subsumed within early *Homo*?

Fossil Evidence

This paper interprets "early" to include fossils which are dated to 1.5 Ma, or before.

This effectively restricts the sample to Africa, for very few hominid sites in Asia can be dated with any reliability to much beyond 1 Ma (Swisher *et al.* 1994; Huang *et al.* 1995). The majority of the African fossil evidence for early *Homo* comes from East Africa, with Olduvai Gorge and sites in the Omo Group, notably Koobi Fora, contributing most of the sample. Specimens from Sterkfontein (*e.g.*, Stw 53) and Swartkrans (*e.g.*, SK 847) also belong in the early *Homo* category, but they will not be referred to in detail in this discussion; the affinities of the latter have recently been reviewed (Grine *et al.* 1993).

Details of the early *Homo* hypodigm have been presented elsewhere (Wood 1991, 1992); the better preserved specimens are listed in Table 1. The material can be divided into two main morphological categories. Some of the remains share features with Asian *H. erectus* and have been referred to in the literature as *H. erectus* (Brown *et al.* 1985), early African *H. erectus* (Bilsborough 1992), or as *Homo ergaster* (Wood 1992). These remains range in age between 1.5 and 1.9 Ma, with most of the sample coming from the younger end of the range. The differences which distinguish them from the remainder of the early *Homo* sample include increases in the overall length of the cranium, the breadth of the cranial vault, and the width of the nasal aperture, reductions in the length of the cranial base and the width of the postcanine teeth, and a posterior attenuation of the molar tooth row (Wood 1991:277).

TABLE 1. Examples of the Remains Attributed to Early *Homo* Listed by Site.

	Koobi Fora (KNM-ER)	*Olduvai* (OH)	*Omo*	*West Turkana* (KNM-WT)
Skulls and crania	730, 1470, 1590, 1805, 1808, 1813, 2592, 3732, 3733, 3735,3883	7, 13, 16, 24, 62	L894-1	15000
Mandibles	730, 820, 992, 1482, 1802, 3734	37, 62	Omo 222-2744	—
Associated teeth	808, 1814	39	—	5496
Postcranial	730, 1472, 1481, 1808, 3228, 3735	7, 8, 35, 62	—	15000

The second of the two main morphological categories includes specimens which have been assigned, or likened, to *H. habilis*. The original hypodigm of *H. habilis* was identified among remains recovered from Beds I and II at Olduvai Gorge (Leakey *et al.* 1964). Since then the hypodigm has been expanded to include other hominid remains from Beds I and II at Olduvai and a substantial sample recovered from the Koobi Fora and Shungura Formations, now known to be components of a larger regional complex of fossiliferous sediments referred to as the Omo Group (Feibel *et al.* 1989). The *H. habilis* category subsumes a substantial range of size and shape variation which has been the subject of different interpretations. Tobias (1991) and Miller (1991), the former on the basis of a wide range of features and the latter on cranial capacity alone, assess this variability to be compatible with an attribution to a single species. Wood (1985), Stringer (1986), Chamberlain (1989), Rightmire

(1993), Kramer *et al.* (1995), and Grine *et al.* (1996) judge the extent and nature of the variation to be less easily accommodated within one species and have proposed that the remains should be attributed to not one, but two species. Stringer (1986) divides the sample temporally, Chamberlain (1989) geographically, and Rightmire (1993) phenetically. Wood (1991, 1992) concurs with Chamberlain's judgement that the Olduvai hypodigm of *H. habilis* is not excessively variable, but does find evidence of excessive variability in the Koobi Fora sample. In Wood's scheme two taxa are recognized at Koobi Fora, one (*H. habilis*) which is also sampled at Olduvai, and the other (*Homo rudolfensis* Alexeev, 1986) which is not. The taxonomic separation Wood proposes was based on the cranial remains, as are all the others cited above, and the cardinal features distinguishing the two taxa are summarized in Table 2.

Why *Homo*?

If we accept the taxonomic interpretation of early *Homo* referred to above, (a subdivision into three species: *H. ergaster*, *H. habilis*, and *H. rudolfensis*), what are the grounds for including these species taxa within the genus *Homo*?

Several morphological trends set *H. erectus* and *H. sapiens* apart from the various species of australopithecine, but five are particularly evident. Firstly, the two species have a larger body size than australopithecine species. Secondly, the brain sizes of fossils assigned to the two species are both absolutely and relatively larger than those of the australopithecines. Thirdly, when compared with australopithecines, for both *H. sapiens* and *H. erectus* there is clear evidence of a reduction in emphasis on the postcanine teeth. This is reflected in both absolute and relative premolar and molar tooth size reduction and in the posterior attenuation of the molar tooth row, so that the third molars, which in the australopithecines are the largest teeth in the molar row, reduce in size so much that in modern humans they are the smallest. Fourthly, there is a shift in locomotor mode such that later *Homo* species are obligatory and not facultative bipeds (Prost 1980). Fifthly and finally, it is one of the characteristics of *Homo* that somatic growth and development are retarded relative to the apes, resulting in the birth of infants which are secondarily altricial.

To what extent do the three early *Homo* taxa conform with these trends? Turning first to *H. ergaster*, there is little doubt that the remains attributed to this taxon provide compelling evidence that, relative to australopithecines, this species demonstrates increases in body and brain size, a reduction in absolute and relative postcanine tooth size, and posterior attenuation of the molar row. Evidence about the rate of maturation in this species is confusing, but observations on the mandible KNM-ER 820 suggest that at least one individual in this group demonstrates a pattern of dental development not unlike that of australopithecines and quite unlike that of later *Homo* species.

The features of the *Homo habilis* sample provide a much more ambiguous message about its taxonomic affinities. If, as seems likely, OH 62 and KNM-ER 3735 prove to be conspecific, then deductions based on these two partial skeletons suggest that *H. habilis* was a creature whose body weight and stature were well within the range of the temporally earlier and morphologically more primitive australopithecine

TABLE 2. Major Differences Between *Homo habilis* and *Homo rudolfensis*

	Homo habilis	Homo rudolfensis
Absolute brain size	$\overline{X}$ = 610 cc	$\overline{X}$= 751 cc
Relative brain size	EQ approx. 4	EQ approx. 4
Overall cranial vault morphology	Enlarged occipital sagittal contribution	Primitive
Endocranial morphology	Primitive sulcal pattern	Frontal lobe asymmetry
Suture pattern	Complex	Simple
Frontal bone	Incipient supraorbital torus	Torus absent
Parietal bone	Coronal > sagittal chord	Primitive
Face - overall	Upper face > midface breadth	Midface > upper face breadth; markedly orthognathic
Nose	Margins sharp, everted; evident nasal sill	Less marginal eversion; no nasal sill
Malar surface	Vertical or near vertical	Anteriorly inclined
Palate	Foreshortened	Large
Upper teeth	Probably two-rooted premolars	Premolars three-rooted; absolutely and relatively large anterior teeth
Mandibular fossa	Relatively deep	Shallow
Foramen magnum	Orientation variable	Anteriorly inclined
Mandibular corpus	Moderate relief on external surface	Marked relief on external surface
	Rounded base	Everted base
Lower teeth	Buccolingually-narrowed postcanine crowns	Broad postcanine crowns
	M_3 reduction	No M_3 reduction
	Reduced talonid on P_4	Relatively large P_4 talonid
	Mostly single-rooted	Twin, plate-like, P_4 roots, and 2T, or even twin, plate-like P_3 roots
Limb proportions	Ape-like	?
Forelimb robusticity	Ape-like	?
Hand	Mosaic of ape-like and modern human-like features	?
Hindfoot	Retains climbing adaptations	?
Femur	Australopithecine-like	?

species. In terms of absolute and relative brain size, *H. habilis* does show an advance with respect to australopithecines. Postcanine tooth area is absolutely small in *H. habilis*, but when scaled by estimated body weight the relative size of premolar and molar crowns is little different than in the so-called "gracile" australopithecine species (Wood 1995). Postcranially the skeleton of OH 62 is primitive with respect to the size and proportion of the skeleton when compared with the later *Homo* species. There is some evidence about the rate and pattern of *H. habilis* development which suggest that it is like the pattern observed in the australopithecines (Dean in press).

The initial inclusion of *H. habilis* within *Homo* was heavily dependent upon interpretations of the OH 8 foot which suggested that it belonged to a hominid with "an upright stance and a fully bipedal gait" (Day & Napier 1964:970). Given that more recent studies have drawn attention to the mosaic nature of the OH 8 foot, with the morphological correlates of prehensility being combined with features linked to bipedalism, do any of the cranial and dental differences between this species and *Australopithecus* still justify its inclusion in *Homo*? The shape and proportions of the cranium and the detailed morphology of the teeth provide the clearest case for separating *H. habilis* from the australopithecines. Although little different in brain capacity, the neurocranium and face of *H. habilis* are proportioned much more like that of later *Homo* crania. The cranial base is also absolutely and relatively shorter, the foramen magnum more centrally-placed and horizontally-inclined, and the occipital bone makes a greater contribution to the sagittal arc length. The midface is no longer the broadest of the three facial components, and the postcanine teeth are relatively narrow buccolingually. The ontogenetic basis and the functional significance of the differences between *H. habilis* and the australopithecines are ill-understood, but they do suggest that the structure of the cranium of *H. habilis* was beginning to be reorganized along the lines of later *Homo*.

There is, as yet, no well-preserved or even partial skeleton of *Homo rudolfensis,* so its limb anatomy can only be inferred from contemporary, but indirectly associated, postcranial remains. Body weight estimates based on orbital dimensions suggest that the body mass of *H. rudolfensis* was on the order of 40-65 kg (Aiello & Wood 1994). The absolute brain size estimates for *H. rudolfensis* are large, but in terms of relative brain size it is comparable to that of *H. habilis*. The locomotor anatomy of *H. rudolfensis*, as inferred from unassociated specimens such as the femora KNM-ER 1471 and 1482, suggests that it was probably an obligatory biped.

It is the cranial and dental anatomy of *H. rudolfensis* which is closest to that of the known australopithecine taxa. The cranium of *H. rudolfensis* is apparently well-pneumatized and its greatest width is across the mastoid region of the temporal bone. The midface is broad, the malar region deep, the face orthognathic, the crowns and roots of the premolar teeth complex, the molars relatively large, and the enamel of the postcanine teeth is relatively thick. All these features point to *H. rudolfensis* retaining a masticatory system that is more australopithecine-like than *Homo*-like.

Is the Genus *Homo* a Monophyletic Group?

A monophyletic group comprises "an ancestral species and all of its descendants" (Wiley *et al.* 1991). Do the species that have been identified as belonging to *Homo* make up what Wiley and his colleagues also call "a unit of evolutionary history"? Very few hominid taxonomic schemes in the contemporary literature identify an ancestral species for the *Homo* clade. In the few that do, *A. africanus* is usually cast in the role of ancestor. However, cladograms which place *A. africanus* as the stem species of *Homo* are only narrowly more parsimonious than schemes which link *A. africanus* with the *Paranthropus* clade (Wood 1991). Others cite *A. africanus*, or a creature very like it, as the common ancestor of a clade incorporating both *Homo* and

Australopithecus (or *Paranthropus*) *boisei* and *robustus* (Skelton & McHenry 1992). If further research supports the claims that *A. africanus* is the exclusive ancestral species for the *Homo* monophyletic group, then the earlier suggestion (Robinson 1965) that *A. africanus* should be abandoned in favor of *H. africanus* should be considered.

The results of cladistic analyses based on cranial, dental, and mandibular characters emphasize the similarities between the *H. rudolfensis* subset of early *Homo* and *Paranthropus* (Wood 1991). Indeed, it is only marginally less parsimonious for *H. rudolfensis* to be the sister taxon of *Paranthropus* than *H. habilis*. However, masticatory morphology is among the less reliable indicators of evolutionary relationships for, even in its finer detail, it is prone to homoplasy in a range of African large mammals (Turner & Wood 1993). Of the two options, the one that regards the "*rudolfensis*" hypodigm as a gnathically-modified, larger-brained, bipedal species of *Paranthropus* is more exotic than the second interpretation, which is that the *Paranthropus*-like morphology of *H. rudolfensis* is either a reflection of the closeness of the relationship between *Homo* and *Paranthropus* (*i.e.*, the morphological complex represents a set of retained ancestral characters; see Skelton & McHenry 1992) or that it reflects similar adaptive responses within the *Paranthropus* and *Homo* clades to external, probably climatic, factors which, in turn, affected the nature of the food available to the hominids (*i.e.*, the morphological complex is homoplasic).

Early *Homo* — One or More Evolutionary Grades?

The three species which are presently subsumed in the informal category early *Homo* represent a wide adaptive range. Early African *H. erectus*, or *H. ergaster*, is the only one of the three species in which overall body size and shape, absolute brain size, masticatory anatomy, and posture and locomotion are all unambiguously tending towards the highly derived conditions we see in later *Homo* species. In many ways early African *erectus* would make a credible ancestral species for a genus restricted to *H. erectus*, *Homo neanderthalensis*, any other species taxa recognized within Middle Pleistocene *Homo*, and *H. sapiens*. In constrast, *H. rudolfensis* retains an australopithecine-grade masticatory system and *H. habilis* an australopithecine-grade physiognomy and locomotor system. The grounds for their inclusion in *Homo* are a good deal more tenuous than those for including early African *H. erectus*.

On the present evidence there are grounds for reconsidering the boundaries of the genus *Homo* and setting them to include the monophyletic group comprising early African *H. erectus* (*H. ergaster*) as the ancestral species and *H. erectus*, *H. sapiens*, and whatever other species taxa should be used to classify *Homo* from the Middle Pleistocene as the remainder.

Conclusions

The considerations discussed above serve to remind us that the fossil record is relatively mute about the events which resulted in the emergence of *Homo*. We are aware from the more recent fossil record of the major morphological "components" of *Homo*, but we remain ignorant about their functional interrelationships, the order in which they arose, and their relationships to the environmental and ecological pressures and constraints prevailing around 2 Ma. Of the approaches presently being explored to remedy this ignorance, the pursuit of the constraints imposed by the need to balance heat, water, and energy budgets (Wheeler 1984, 1985, 1991a and b, 1992a and b, 1993; Ruff 1991) and studies of the evolutionary history of contemporary large-bodied mammals (*e.g.*, Bishop 1994) hold out the greatest promise.

Acknowledgements

Research incorporated in this paper was supported by grants provided by The Leverhulme Trust and the Science-based Archaeology Committee of the SERC.

Literature Cited

Aiello, L. C., & B. A. Wood. 1994. Cranial variables as predictors of hominine body mass. *American Jour. Phys. Anthropol.* 95:409-426.

Bilsborough, A. 1992. *Human Evolution.* Blackie, London, UK. 258 pp.

Bishop, L. C. 1994. *Pigs and the Ancestors: Hominids, Suids and Environments during the Plio-Pleistocene of East Africa.* Ph.D. Dissertation, Yale University, New Haven, Connecticut.

Brown, F., J. Harris, R. Leakey, & A. Walker. 1985. Early *Homo erectus* skeleton from west Lake Turkana, Kenya. *Nature* 316:788-792.

Chamberlain, A. T. 1989. Variations within *Homo habilis.* Pages 175-181 *in* G. Giacobini, ed., *Hominidae.* Jaca Books, Milan, Italy.

Day, M. H., & J. R. Napier. 1964. Fossil foot bones. *Nature* 201:969-970.

Dean, M. C. (in press). The nature and periodicity of incremental lines in primate dentine and their relationships to periradicular bands in OH 16 (*Homo habilis*). *In* J. Moggi Cecchi, ed., *Aspects of Dental Biology: Palaeontology, Anthropology and Evolution.* A. Angelo Pontecorboli, Florence, Italy.

Feibel, C. S., F. H. Brown, & I. McDougall. 1989. Stratigraphic context of fossil hominids from the Omo Group deposits: Northern Turkana Basin, Kenya and Ethiopia. *American Jour. Phys. Anthropol.* 78:595-622.

Grine, F. E. 1993. Australopithecine taxonomy and phylogeny: Historical background and recent interpretation. Pages 198-210 *in* R. L. Ciochon & J. G. Fleagle, eds., *The Human Evolution Source Book.* Prentice Hall, Englewood Cliffs, New Jersey.

_____ , B. Demes, W. L. Jungers, & T. M. Cole III. 1993. Taxonomic affinity of the early *Homo* cranium from Swartkrans, South Africa. *American Jour. Phys. Anthropol.* 92:411-426.

_____ , W. L. Jungers, & J. Schultz. 1996. Phenetic affinities among early *Homo* crania from East and South Africa. *Jour. Hum. Evol.* 30:189-225.

Huang, W., R. Ciochon, Y. Gu, R. Larick, Q. Fang, H. Schwarcz, C. Yonge, J. de Vos, & W. Rink. 1995. Early *Homo* and associated artefacts from Asia. *Nature* 378:275-278.

Kramer, A., S. M. Donnelly, J. H. Kidder, S. D. Ousley, & S. M. Olah. 1995. Craniometric variation in large-bodied hominoids: Testing the single-species hypothesis for *Homo habilis*. *Jour. Hum. Evol.* 29:443-462.

Leakey, L. S. B., P. V. Tobias, & J. R. Napier. 1964. A new species of the genus *Homo* from Olduvai Gorge. *Nature* 202:7-9.

Mayr, E. 1950. Taxonomic categories in fossil hominids. *Cold Spring Harbor Symp. Quant. Biol.* 15:109-118.

Miller, J. A. 1991. Does brain size variability provide evidence for multiple species in *Homo habilis*? *American Jour. Phys. Anthropol.* 84:385-398.

Prost, J. H. 1980. Origin of bipedalism. *American Jour. Phys. Anthropol.* 52:175-189.

Rightmire, G. P. 1993. Variation among early *Homo* crania from Olduvai Gorge and the Koobi Fora region. *American Jour. Phys. Anthropol.* 90:1-33.

Robinson, J. T. 1965. *Homo 'habilis'* and the australopithecines. *Nature* 205:121-124.

_____. 1972. The bearing of East Rudolf fossils on early hominid systematics. *Nature* 240:239-240.

Ruff, C. B. 1991. Climate, body size and body shape in hominid evolution. *Jour. Hum. Evol.* 21:81-105.

Simpson, G. G. 1961. *Principles of Animal Taxonomy*. Columbia University Press, New York. 247 pp.

Skelton, R. R., & H. M. McHenry. 1992. Evolutionary relationships among early hominids. *Jour. Hum. Evol.* 23:309-349.

Stringer, C. B. 1986. The credibility of *Homo habilis*. Pages 266-294 *in* B. Wood, L. Martin, & P. Andrews, eds., *Major Topics in Primate and Human Evolution*. Cambridge University Press, Cambridge, UK.

Swisher, C. C., G. H. Curtis, T. Jacob, A. G. Getty, A. Suprito, & Widiasmoro. 1994. Age of the earliest known hominids in Java, Indonesia. *Science* 263:1118-1121.

Tobias, P. V. 1991. *Olduvai Gorge, Volume 4: The Skulls, Endocasts and Teeth of* Homo habilis. Cambridge University Press, Cambridge, UK. 921pp.

Turner, A., & B. A, Wood. 1993. Comparative palaeontological context for the evolution of the early hominid masticatory system. *Jour. Hum. Evol.* 24:301-318.

Wheeler, P. E. 1984. The evolution of bipedality and loss of functional body hair in hominids. *Jour. Hum. Evol.* 13:91-98.

_____. 1985. The loss of functional body hair in man; the influence of thermal environment, body form and bipedality. *Jour. Hum. Evol.* 14:23-28.

_____. 1991a. The thermoregulatory advantages of hominid bipedalism in open equatorial environments: The contribution of increased convective heat loss and cutaneous evaporative cooling. *Jour. Hum. Evol.* 21:107-115.

_____. 1991b. The influence of bipedalism on the energy and water budgets of early hominids. *Jour. Hum. Evol.* 21:117-136.

_____. 1992a. The influence of the loss of functional body hair on the energy and water budgets of early hominids. *Jour. Hum. Evol.* 23:379-388.

_____. 1992b. The thermoregulatory advantages of large body size for hominids foraging in savannah environment. *Jour. Hum. Evol.* 23:351-362.

_____. 1993. The influence of stature and body form on hominid energy and water budgets; A comparison of *Australopithecus* and early *Homo* physiques. *Jour. Hum. Evol.* 24:13.

Wiley, E. O., D. Siegel-Causey, D. R. Brooks, & V. A. Funk. 1991. *The Compleat Cladist: A Primer of Phylogenetic Procedures*. University of Kansas Museum of Natural History, Lawrence, Kansas. 158 pp.

Wood, B. A. 1985. Early *Homo* in Kenya and its systematic relationships. Pages 206-214 *in*
    E. Delson, ed., *Ancestors: The Hard Evidence.* Alan R. Liss, New York.

_____. 1991. *Koobi Fora Research Project. Vol. 4: Hominid Cranial Remains.* Clarendon
    Press, Oxford, UK. 466 pp.

_____. 1992. Origin and evolution of the genus *Homo. Nature* 355:783-790.

_____. 1995. Evolution of the early hominin masticatory system: Mechanisms, events and
    triggers. Pages 438-448 *in* E. S. Vrba *et al.*, eds., *Paleoclimate and Evolution with Emphasis
    on Human Origins.* Yale University Press, New Haven, Connecticut.

Current Issues in Modern Human Origins

Christopher B. Stringer
Department of Palaeontology
The Natural History Museum
Cromwell Road
London SW7 5BD, UK

The study of modern human origins is marked by fundamental disagreements about the best methods for recognizing and analyzing the variation among fossil samples. Distinguishing intraspecific from interspecific variation is crucial for classifying Pleistocene *Homo*. Four recent proposals relevant to understanding modern human origins are discussed: that recognition of only three species of the genus *Homo* underestimates the true number of species; that *Homo erectus* should be sunk into *Homo sapiens*; that all Middle Paleolithic human fossils from the Levant represent variants of a single population; that the Skhul-Qafzeh samples are highly variable and have been mischaracterized as "modern." The most reasonable division of middle and later Pleistocene *Homo* is into at least four species: *H. erectus, H. neanderthalensis, H. sapiens*, and a common ancestor to *H. sapiens* and *H. neanderthalensis*. This fourth species may be referred to as *H. heidelbergensis, H. rhodesiensis*, or *H. helmei*, depending on its precise composition.

The recent ferment in the study of modern human origins has led to an unprecedented burst of scientific activity in the form of fieldwork and research, in debate, both in the scientific and popular media, and in a reexamination of fundamentals in the subject. In this paper I intend to look critically at three basic issues and discuss some recent proposals concerning research in modern human origins. The first two issues are absolutely fundamental in that they concern the way we group and study the fossil material. The most widely used taxonomy of the genus *Homo* recognizes the existence of three species, *Homo habilis, Homo erectus*, and *Homo sapiens*. *Homo habilis* was restricted to Africa and was apparently extinct by 1.5 Ma. The taxon *Homo erectus* is known from Africa and Asia and usually includes African forms dating prior to one million years (for example Olduvai Hominid 9 and the Koobi Fora KNM-ER 3733 specimen). *Homo sapiens* is usually informally divided into an "archaic" subdivision, including most African and European Middle Pleistocene fossils as well as the Late Pleistocene Neanderthals, and an "anatomically modern" subdivision comprising forms such as those from Skhul, Qafzeh, and the European Upper Paleolithic, as well as recent human samples.

Proposal 1: Recognition of only three species of the genus *Homo* underestimates the true number of species

Tattersall (1986, 1992) has commented that living primate species numbers would be seriously underestimated if we had to rely on only the skeletal parts which might be preserved as fossils. He has therefore suggested that species numbers in fossil hominids are probably underestimated, and that the Neanderthals should be distinguished as a separate species, with the likely addition of several more species requiring recognition in the Early and Middle Pleistocene.

As we examine and interpret the fossil record to reconstruct phylogenies, we require a reasonable basis for species recognition which will allow a clear demarcation between intraspecific and interspecific variation. Otherwise, there is the likelihood that significant taxonomic distinctions will be trivialized as intraspecific polymorphisms, leading to a failure to recognize significant speciation events, or conversely, that relatively trivial intraspecific differences will be elevated to a level of fundamental significance. The best known species concept is that of the "Biological Species," which refers to the presence of a common gene pool. This concept is necessarily only of direct applicability to the living biota, but many workers have attempted to use skeletal variation present within living species as a yardstick by which to judge past specific variation. However, there are immediate problems here for the hominid fossil record when we consider whether to use only present-day human variation as a guideline or to extend comparisons to include other primates. As we have seen from Proposal 1, features which separate known, distinct living primates species may be difficult or impossible to find in the skeleton. Another species concept which is difficult to employ in the fossil record, whatever its theoretical validity, is the "Recognition Concept." This argues that the primary cause of speciation is not the separation of once common gene pools by isolation, but a divergence in the once shared Specific Mate Recognition System (SMRS) within a species. The SMRS can range from a biochemical signal, such as a particular pheromone, to an elaborate morphological display, such as the shape of antlers common to the males of a particular species of deer. A change in the SMRS in some members of the species can produce an immediate behavioral or even physical barrier to fertilization with other members of the species. According to the Recognition Concept, most of the morphological features which we can observe as separating related species have developed independently of rifts in previously shared fertilization systems.

Because of the inherent difficulties in relating species concepts based on living species to the fossil record, other workers have preferred to employ concepts based on the presence of synapomorphies (shared derived characters). For those who are prepared to go further than what has been termed "transformed cladistics" or "natural order systematics," the presence of such shared characters form the basis of a

"Phylogenetic Species Concept," which arises from the evolutionary process: a species is the basal group of organisms within which there is a parental pattern of ancestry and descent. Some workers have favored the use of autapomorphies in species recognition (for example Tattersall 1986). Others are prepared to use combinations of apomorphies and plesiomorphies as long as the combination is unique to the species concerned. A related, but not necessarily cladistically based, concept is that of the "Evolutionary Species," where branching points determine species recognition, and the unique evolutionary role of a species is the basis of its recognition. In an important volume (Kimbel & Martin 1993), the use of these various species concepts is reviewed, and further proposals regarding species recognition are made. In their contribution to the Kimbel & Martin volume, Kimbel & Rak (1993) support the use of the Phylogenetic Species Concept and make the point that it does not have to be based on autapomorphies alone. A species can be recognized by any unique set of character states, and prior knowledge of the cladistic status of the characters and the species is not required — it only needs to be diagnosable in space and time. Harrison (1993) goes even further in his discussion of cladistics and the species problem. Although a confirmed cladist, Harrison argues that species diagnosis need have no cladistic basis — it is merely a phenetically determined distinctive morphotype. A combination definition of characters which may be known to be, or may turn out to be, apomorphies or plesiomorphies provides a practical approach to species recognition, and a differential diagnosis (distinguishing species individually from other species) is preferable to a single universal diagnosis.

Tattersall (this volume; see also Tattersall 1992) has also moved away from his earlier emphasis on autapomorphies to a position where apomorphies are employed as part of "morph" recognition. Given the lack of morphological differentiation between many closely related primate species, Tattersall argues that if discrete morphs (minimum diagnosable units) can be recognized in fossil assemblages, these probably do represent distinct species. This practical, rather than theoretical, proposal is one I also favor for species recognition in the fossil record. Tattersall argues that good diagnosable characters for (anatomically modern) *Homo sapiens, Homo neanderthalensis*, and *Homo heidelbergensis* exist, amongst the forms generally recognized as *Homo sapiens* (*sensu lato*), and that these should all be recognized specifically. However, he also claims that no reasonable list of shared characters has ever been produced for *Homo sapiens* (*s.l.*). In fact I did produce one (Stringer 1984) as part of a discussion of the status of the species *Homo erectus* (Table 1), and although I no longer consider this list appropriate for a species diagnosis, I will return to it later.

Proposal 2: *Homo erectus* should be sunk into *Homo sapiens*

A contrasting proposal which has been revived recently in conjunction with discussion of the multiregional model of modern human origins is that the species *Homo erectus* should be sunk into *Homo sapiens* (Wolpoff *et al.* 1994), with *Homo sapiens* originating in a late Pliocene cladogenetic event. This sinking of the species

TABLE 1. Possible synapomorphies of *Homo sapiens* (*sensu lato*)
(vs. *H. erectus*), modified from Stringer (1984)

1. More gracile tympanic
2. Greater midface projection
3. Laterally retracted supraorbital torus
4. Longer, more curved parietals
5. Higher, more curved temporal squama
6. Lengthened occipital plane
7. Larger endocranial volume
8. Reduced total facial prognathism
9. Parallel-sided or expanded parietal arch
10. Reduced cranial cortical bone
11. Reduced occipital torus
12. Laterally reduced supraorbital torus
13. No iliac pillar

Homo erectus is justified by the claim for genetic continuity between Early and Late Pleistocene human populations in the various regions of the Old World, for example between Neanderthals and modern humans in Western Eurasia; between populations represented by the Early Pleistocene Lantian and Middle Pleistocene Zhoukoudian fossil samples and recent oriental and native American populations; and between the Indonesian samples known from sites such as Sangiran and Ngandong, and native Australian populations.

Wolpoff *et al.* (1994) argue that *Homo sapiens* can be viewed as a single evolutionary species spanning the entire Pleistocene. "Evolutionary tendencies" are linked to the evolving cultural system and geological criteria are proposed for early, middle, and late *sapiens* categories, apparently correlated to the Early, Middle, and Late Pleistocene. If the multiregional model of modern human origins is actually representative of the course of human evolution over the last million years or so, then there is no doubt that only one species of human should be recognized (*Homo sapiens*). However, the Wolpoff *et al.* proposal to sink *Homo erectus* is, like all species assignments and phylogenies employing fossils, an hypothesis about what really happened in the past. As such it should be testable, although the authors have provided little or no basis for testing. Instead it is assumed that the relationships are there and merely need to be supported by data collection (Stringer & Bräuer 1994). Certainly the validity of the proposed cladogenetic origin of *Homo sapiens* (*sensu lato*) can be tested to see whether it corresponds to an identifiable speciation event, and most workers would agree that there is good evidence for cladogenesis with the origin of *Homo erectus* (*sensu lato*) (or *Homo ergaster* of Wood 1992), which would then mark the origin of the *Homo sapiens* (*sensu lato*) clade of Wolpoff *et al.* (1994).

However, their proposal that the polytypism of present day *Homo sapiens* is comparable with, and an extension of, the polytypism of their *Homo sapiens* evolutionary species throughout the Pleistocene seems difficult to accept. Unfortunately we lack large, contemporaneous samples for Pleistocene hominids (with the probable exception of Atapuerca, discussed below). However, we do have good data on modern morphological polytypism. Table 2a presents data for cranial samples of four regionally distinct recent groups with an additional contrast provided by sexual dimorphism. The measurements are those of basion-nasion length (BNL), midline nasal height (NLH), midline projection of the lower nasal margin (SSS), frontal doming (FRS), parietal doming (PAS), and position of midline maximum occipital projection (OCF). For comparison, the means for five later Pleistocene cranial samples (which would all be included in Wolpoff et al.'s (1994) "late sapiens" category) are provided. The mean data show considerable variation, but much of this is size related. If we standardize for size by log transformation and row standardization (Corruccini 1987), we are left with a more realistic picture of shape differences among the recent cranial samples, and these are actually very small indeed (Table 2b). By contrast, the transformed data for the fossil samples show greater variation, indicating that the distinctions among the fossil groups are of a different order from those among the recent samples — a point which has been made many times before (e.g., Howells 1989; Stringer 1974; van Vark & Bilsborough 1991). Of course, the fossil samples probably span something like 100,000 years, while the recent samples (although covering a greater geographical range) are penecontemporaneous, yet we can immediately reduce the between-sample differences in the fossils by regrouping the data as shown in Table 2c. Here the division corresponds to what I would regard as modern vs. non-modern, and the time span of the "modern" samples is now comparable with the much more varied fossil samples of Table 2 — about 100,000 years. Even now, despite the reduction of the archaic among-sample differences, the metrical differences are still more marked in the archaic than the more widely-based "modern" grouping.

Thus, there appears to be little or no justification for the view that modern variation is equivalent to variation at earlier time levels in the Pleistocene. Another argument used to support the merging of Homo erectus and Homo sapiens is the claim for continuity of characters between the two species in the same regions. The recent proponents of regional continuity have never used global skeletal samples to provide data in support of their ideas, and an examination of such data gives poor support for regional continuity (see for example Howells 1973, 1989; Corruccini 1992; Groves 1989; Stringer 1992a and b; Lahr 1994). The fundamental mechanism for the maintenance of relatively minor regional characters over a period of one million years or so in the face of major evolutionary changes in the cranium (not to mention major and repeated climatic changes) is also unclear. While drift and selection were formerly proposed as local mechanisms of maintenance (Wolpoff et al. 1984; Wolpoff 1989) the role of selection is now being played down (Thorne & Wolpoff 1992), leaving an explanatory void.

The primary test of regional continuity is to establish whether it operates in the Late Pleistocene. If it does not operate during the last stages of human evolution, then the regional sequences are broken, however convincing they look at an earlier date in areas such as Europe and Indonesia, where most workers do recognize regional continuity in the Middle Pleistocene. Because early modern fossils in Europe, Asia, and Africa do not closely resemble their modern counterparts phenetically or in

TABLE 2. "Modern" vs. "Archaic" cranial data comparisons

NORSEM, DOGONF, AUSM and JAPF are respective male (M) or female (F) means of the Howells samples from Norway, West Africa, Australia, and Japan. EUNEAM and ASNEAM are means of late Neanderthal samples from Europe and Asia, AFLARCH is the African late archaic mean, SQMEAN is Skhul-Qafzeh and EUPMEAN early Upper Paleolithic mean.

	BNL	NLH	SSS	FRS	PAS	OCF
2a. RAW DATA (mm)						
NORSEM	101.80	51.96	22.96	25.11	24.69	47.40
DOGONF	94.87	46.17	21.04	25.67	22.25	44.38
AUSM	101.98	49.69	24.13	25.23	23.90	43.52
JAPF	95.63	49.37	21.73	25.59	23.49	46.83
EUNEAM	113.50	60.64	38.00	20.72	17.56	39.75
ASNEAM	116.40	64.93	38.00	19.07	21.97	45.40
AFLARCH	105.33	47.25	29.50	23.33	17.50	43.33
SQMEAN	102.88	52.67	25.50	25.30	22.70	52.50
EUPMEAN	101.13	50.79	24.17	28.14	23.37	49.50
2b. STANDARDIZED DATA						
STNORSEM	0.96	0.28	-0.53	-0.44	-0.46	0.19
STDOGONF	0.96	0.24	-0.55	-0.35	-0.49	0.20
STAUSM	0.98	0.26	-0.46	-0.42	-0.47	0.13
STJAPF	0.93	0.27	-0.55	-0.39	-0.47	0.22
MAXDIFF	0.05	0.04	0.09	0.09	0.03	0.09
STENEA	1.06	0.43	-0.04	-0.64	-0.81	0.01
STASNEA	1.02	0.44	-0.10	-0.79	-0.65	0.08
STAFARCH	1.04	0.24	-0.23	-0.46	-0.75	0.16
STSQ	0.94	0.27	-0.45	-0.46	-0.57	0.27
STEUP	0.93	0.24	-0.50	-0.35	-0.54	0.22
MAXDIFF	0.13	0.20	0.46	0.44	0.27	0.26
2c. REARRANGED DATA						
STNORSEM	0.96	0.28	-0.53	-0.44	-0.46	0.19
STDOGONF	0.96	0.24	-0.55	-0.35	-0.49	0.20
STAUSM	0.98	0.26	-0.46	-0.42	-0.47	0.13
STJAPF	0.93	0.27	-0.55	-0.39	-0.47	0.22
STSQ	0.94	0.27	-0.45	-0.46	-0.57	0.27
STEUP	0.93	0.24	-0.50	-0.35	-0.54	0.22
MAXDIFF	0.05	0.04	0.10	0.11	0.11	0.14
STENEA	1.06	0.43	-0.04	-0.64	-0.81	0.01
STASNEA	1.02	0.44	-0.10	-0.79	-0.65	0.08
STAFARCH	1.04	0.24	-0.23	-0.46	-0.75	0.16
MAXDIFF	0.04	0.20	0.19	0.33	0.16	0.15

discrete morphological characters, either they are not ancestral to those counterparts, or regionality mainly evolved in the recent past (see for example Groves 1989; Stringer 1992a and b; Wright 1992). Some (but not all) Australian fossils do show modern regional characters, but I have suggested that this is because many of these "Australian" characters were widespread in the Late Pleistocene, so that fossils in Africa, Europe, and Asia also look "Australian" (Stringer 1992a and b). Even where more carefully argued cases for morphological continuity between late archaic and early modern specimens have been made (for example Smith & Trinkaus 1991;

Frayer 1992) the characters are not generally those which are otherwise claimed to be long-term regional markers. This suggests that the regionality of modern humans is quite different from, and quite independent of, the regionality of premodern humans.

If the proposal to place all hominids of the last million years or more in *Homo sapiens* is not accepted, how can we best partition the fossil samples in relation to modern humans and to each other? I have been quoted as denying that *Homo erectus* was ancestral to *Homo sapiens* (for example Pope 1992:246), but my views have probably been confused with those of a colleague (Andrews 1984). In fact, my view, based on morphology and our present chronological framework, has not changed for many years and is identical with that expressed by Harrison (1993), who otherwise takes issue with the way that Andrews, Wood, and I approached "defining" the species *Homo erectus* in 1984. Harrison states "if we exclude the possibility that all *Homo erectus* populations across the Old World graded imperceptibly into *Homo sapiens* . . ., the presence of unique specializations in the Asian population may indeed make it more likely that the African population is the ancestral population from which *Homo sapiens* was derived."

What Andrews, Wood, and I attempted to do, for the first time in detail, was to see whether *Homo erectus* could be diagnosed by apomorphies, and especially by autapomorphies. My view was that this could not be achieved if the *Homo erectus* sample was taken very widely, but it was more successful if restricted to the Asian hypodigm. However, I recognized that *erectus*-like features existed outside of the Far East, in African fossils such as Olduvai Hominid 9, Bodo, and Omo-Kibish 2, and in European ones such as Petralona, and I recognized that these indicated that an evolutionary relationship probably existed between *Homo erectus* and "archaic *Homo sapiens*" in the western Old World, even if such a relationship did not exist in the East. The strict cladistic approach of the 1984 papers, emphasizing autapomorphies in search of a universal diagnosis of *Homo erectus*, is not something I would advocate now, since I agree that we should attempt to recognize distinct fossil morphs (= paleospecies) not by unique autapomorphies, but by unique combinations of characters (whether plesiomorphies or apomorphies) compared with other morphs.

Having re-established that point, I would like to turn to the question of how best to deal taxonomically with the range of variation in Pleistocene and recent *Homo*. Here I do not want to discuss the question of the status of the earlier African fossils such as KNM-ER 3733 and 3883 attributed by most workers to *Homo erectus* (for example Rightmire 1990), by others to *Homo leakeyi* (Clarke 1990), and by yet others to *Homo ergaster* (Wood 1992). I consider this issue to be unresolved from present data.

Treating the early African specimens KNM-ER 3733, 3883, and Olduvai Hominid 9 as representing a single distinct morph (AFER), I have compared this group with samples of Middle Pleistocene hominids from Africa (Broken Hill, Bodo, Ndutu, Saldanha, Salé – AFMP) and Europe (Petralona, Arago, Steinheim, Bilzingsleben, Vértesszöllös, Swanscombe, Ehringsdorf, Biache – EMP); late Middle to early Late Pleistocene hominids of Africa (sample as Stringer 1992b, except for the parietal data for Singa, omitted due to probable pathology – AFA); European Neanderthals (as Stringer 1992b – ENEA); Skhul-Qafzeh (as Stringer 1992b – SQ); and a European early Upper Paleolithic sample (as Stringer 1992b – UP). Logarithm and row stand-ardized data for 39 measurements have been compared, as well as Penrose size and shape analyses using 35 of these measurements (removing one of each pair of the most highly correlated, as Stringer 1992b). The raw and Penrose size data show that there have been significant increases from AFER to the Late Pleistocene hominids in size, particularly for the EMP, AFMP, ENEA and AFA groups. However, shape differences have also been extensive, and by no means all in the same direction.

Figure 1 shows a phenogram with the square root of the three largest shape distances plotted to scale as a triangle, and the others given their approximate relative positions. If we now plot the shape distances of the groups graphically, using facial data (15 measurements) and vault data (20 measurements) and taking AFER as the base point, Figure 2 is produced. There is certainly no linear trend in shape changes, and instead two main "trends" are evident. The ENEA, AFMP, and EMP groups show greater changes from AFER in face shape than vault shape, while the SQ and UP groups marginally show the opposite. AFA falls into an intermediate position. Looked at from the perspective of the UP group, a much more linear arrangement is produced with more equal face and vault distances (Figure 3). However, taking the overall shape distances for the 35 measurements combined and plotting them against estimated mean age (omitting AFER which has a comparable shape distance from UP to AFMP, but a much greater time difference) produces Figure 4. The discrepancy in the position of the Neanderthals is notable — they are almost certainly closer in mean age to UP than either the SQ or AFA groups, yet they are much more distant in shape. In fact they are at about the same shape distance from UP as the much older EMP sample. Thus there is no straightforward progressive change leading to modern humans of the kind advocated by Wolpoff *et al.* (1994) (nor, incidentally, is there

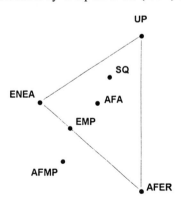

FIGURE 1. Phenogram showing shape distances between African early *Homo erectus* (AFER), European Neanderthal (ENEA), and European Upper Paleolithic (UP) cranial sam-ples (35 measurements). Approximate relative positions of African Middle Pleistocene (AFMP), European Middle Pleistocene (EMP), African late archaic (AFA), and Skhul-Qafzeh (SQ) samples added.

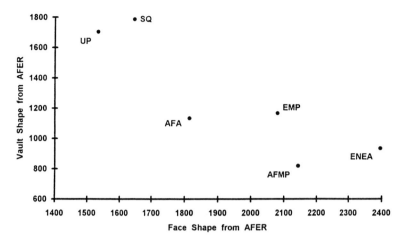

FIGURE 2. Shape distances from AFER group for facial (15 measures) and vault (20 measures) analyses.

evidence for regional continuity in the earlier samples, since I could have substituted any of Howells' modern geographical samples for the UP sample, with little effect on the overall pattern of results).

I am happy with the present evidence for the discreteness of the taxon *Homo erectus* from both Neanderthals and modern *Homo sapiens*, for example using its suite of characters which contrast with those of Table 1. In my view the evidence for the distinctiveness of a Neanderthal morph and Neanderthal lineage is almost as impressive as the distinction between *Homo erectus* and modern *Homo sapiens*. The combination of a highly derived facial morphology with a relatively primitive, but still distinctive cranial vault, and the postcranial combination of probable primitive

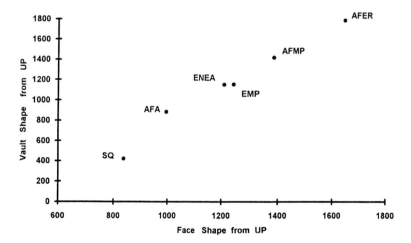

FIGURE 3. Shape distances from UP group for facial and vault analyses.

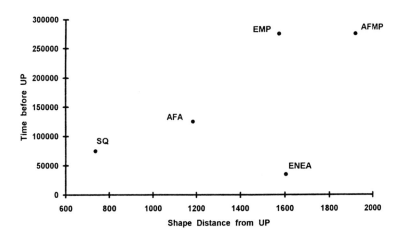

FIGURE 4. Overall shape distances from UP group versus estimated time difference (AFER omitted).

skeletal characters (for example in the anterior pelvis) with a derived body shape paralleling that of modern cold-adapted peoples (see for example Trinkaus 1981; Ruff 1991), marks the Neanderthals as a separate morph well worthy of specific recognition (see also Rak 1993). The proposed resurrection of *Homo neanderthalensis* King, 1864 has gained a number of supporters recently, including Harrison (1993), who is otherwise rather cautious about the proliferation of species names in hominids. He states "I would have to concur with Tattersall that if I found two morphs as different as Neanderthals and modern humans in the Miocene, there is little doubt that I would recognize them as two distinct species."

I have previously argued, like Rightmire (1990), for the existence of a relatively primitive Eur-African Middle Pleistocene hominid subspecies or species which lies close to the common ancestry of Neanderthals and modern humans. This hominid was probably contemporaneous with, but distinct from, Asian *Homo erectus*, confirming that cladogenesis had occurred. If we regard the Late Pleistocene morphs as worthy of specific recognition as *Homo neanderthalensis* and *Homo sapiens*, the specific name *Homo heidelbergensis* appears to be the most appropriate one for the putative ancestral species. Referring back to Table 1, the characters of *Homo sapiens* (*sensu lato*) could be made appropriate for *Homo heidelbergensis* by delineating characters 1-5 as synapomorphies of the taxon shared with the Neanderthal and modern human species, characters 6-11 as characters intermediate between those of *Homo erectus* and later humans, and characters 12 and 13 as more advanced characters lacking in the taxon.

However, recent research and discoveries have thrown the status of *Homo heidelbergensis* into some doubt (Arsuaga *et al.* 1993; Stringer 1993). As we have seen from the previous phenetic analyses, AFMP and EMP (which might be assigned to *Homo heidelbergensis*) predominately resemble *Homo neanderthalensis*, rather than

Homo sapiens, because data for AFMP and EMP are dominated by specimens such as Broken Hill, Bodo, Petralona, and Arago, which share plesiomorphies in vault form and, arguably, some apomorphies in facial form with *Homo neanderthalensis*. Rak (1993) recognizes the uncertain status of these specimens by referring to them as the "Kabwe (= Broken Hill) group." The large series of cranial fossils from Atapuerca has also shown up the problems of recognizing a separate European species prior to the Neanderthals. I had previously differed from Wolpoff (1980) in arguing that the contrasts between the Steinheim and Petralona crania warranted a degree of taxonomic separation between them, rather than merely reflecting within-population variation, especially dimorphism (Stringer 1981, 1985). However, the Atapuerca material includes specimens resembling both Petralona and Steinheim. Yet despite non-Neanderthal details such as a rather modern-looking (plesiomorphous?) temporal bone and parietal region, there are facial, occipital, and mandibular aspects which clearly foreshadow those of the Neanderthals. If the Atapuerca sample is really penecontemporaneous, it probably represents a primitive form of *Homo neanderthalensis,* if we recognize the reality of that taxon. To then maintain that Petralona represents a European species, *Homo heidelbergensis*, distinct from Atapuerca now seems more problematic. Perhaps more complete material of specimens like those from Mauer and Bilzingsleben might show further consistent differences from the Atapuerca range (*e.g.*, in the occipital region and postcrania) which would allow the maintenance of a distinct species, but this remains to be seen.

The situation in China and Africa is more complex still. The zygomaxillary morphology of Broken Hill 1 and Bodo resembles that of Neanderthals (and Petralona) in some respects, whereas that of other African specimens such as Ndutu, Thomas Quarry, Broken Hill 2, Florisbad, Ngaloba L.H. 18 and Jebel Irhoud 1 does not. This could represent polytypism in the African Middle to early Late Pleistocene, either locally developed or alternatively related to gene flow from another region such as the Far East (Li & Etler 1992; Pope 1992). However, because the preservation of the relevant regions in some of the Chinese material is by no means perfect, it is unclear how much variation there really is in the East, and how distinct this is from the pattern in the West. It is also uncertain how the African and Asian specimens precisely relate to each other chronologically. If we were to link what Rak (1993) calls the "Kabwe group" with *Homo heidelbergensis* through facial similarities, there would remain a group of archaic African and Asian specimens which display a zygomaxillary morphology more like that of modern *Homo sapiens* (although it should be cautioned that the relevant morphology of *Homo sapiens* is by no means as consistent as sometimes claimed). The (plesiomorphous ?) facial features of these specimens would serve to distinguish them from *Homo neanderthalensis* and possibly relate them to *Homo sapiens*, while there would also be some plesiomorphous vault features found in early *Homo neanderthalensis* (*e.g.*, Atapuerca). Some of the specimens also show vault features which could be considered synapomorphous with *Homo sapiens*, but there is considerable morphological variation even within the Chinese and African samples, let alone when they are combined.

If a species were to be recognized which was distinctive from both *Homo neanderthalensis* and *Homo sapiens*, and in Europe the name *Homo heidelbergensis* was at the same time sunk into *Homo neanderthalensis*, two other nomina might be appropriate. One would be *Homo rhodesiensis*, if the type specimen from Broken Hill is judged to be distinct from the Neanderthal and modern human clades. Another available name would be *Homo helmei* (Dreyer 1935), based on the Florisbad skull. *Homo helmei* would then include African specimens such as Florisbad, Eliye Springs, Ngaloba, and Irhoud, and probably also Asian specimens such as Dali, Jinniushan, and, perhaps, Maba, Narmada, and Zuttiyeh. Yet another alternative, recently proposed by me (Stringer 1992c), would be to extend the taxon name *Homo neanderthalensis* to encompass all the European, African, and Asian forms previously subsumed under the term "archaic *Homo sapiens*." This proposal would downgrade Neanderthal derived features of the face, vault, and body proportions to the level of "racial variation," and would emphasize shared vault and postcranial features which were mainly plesiomorphous retentions from *Homo erectus* (see, for example, Trinkaus, in press). I am not so happy with this proposal now, because I feel that the Atapuerca finds and other unpublished research reinforce the special nature of Neanderthal evolution in Europe, and this should be recognized taxonomically for continuing research purposes. And that last statement is perhaps the strongest reason for also resisting the proposal to sink *Homo erectus* into *Homo sapiens*. Such a proposal can be the legitimate goal of a research program, and might be vindicated in the long run, but we are a long way from establishing it. The premature extension of the term *Homo sapiens* beyond the Late Pleistocene to the earlier Pleistocene, and even Pliocene, would sound the death knell for meaningful systematic research on Pleistocene hominids, and that is surely something that most of us do not wish to see.

The most useful phylogenetically-based taxonomy for Middle-Late Pleistocene hominids would seem to be the recognition of at least four species: *Homo erectus* (possibly continuing into the late Middle Pleistocene in eastern Eurasia and perhaps even into the Late Pleistocene in Java); *Homo neanderthalensis* (Late Pleistocene of western Eurasia, extending back to the Middle Pleistocene *circa* 200-300 Ka in Europe, based on the Atapuerca evidence, or even 400 Ka based on Swanscombe); *Homo sapiens* (Late Pleistocene, all regions); and a fourth species (Middle Pleistocene, possibly extending into the Late Pleistocene, in Africa and eastern Eurasia and perhaps Europe). Depending on whether holotypes are regarded as validly distinct from the three species already named, the order of priority of nomina for this species is: (i) *Homo heidelbergensis*; (ii) *Homo rhodesiensis*; (iii) *Homo helmei*.

Proposal 3: All Middle Paleolithic human fossils from the Levant represent variants of a single population (and therefore species)
Proposal 4: The Skhul-Qafzeh samples are highly variable and have been mischaracterized as "modern"

These proposals concern the classification of human fossils associated with the Middle Paleolithic in the Levant (that is, the area adjoining the eastern Mediterranean). There is growing evidence that Neanderthal and anatomically modern populations coexisted or alternated in their occupation of the region. However, it has been argued that these are merely variants of a single Middle Paleolithic human population (Wolpoff 1992) and that the Skhul-Qafzeh group is phenetically intermediate between Neanderthals and modern humans (Corruccini 1992).

The Levantine Middle Paleolithic fossil hominid record has long been controversial (see for example McCown & Keith 1939; Howell 1957; Vandermeersch 1981; Trinkaus 1984). Following new discoveries, further studies, and new radiometric dating, the situation had seemed clearer, although not necessarily simpler in its implications, with the recognition of two distinct populations — early modern and Neanderthal — probably overlapping in time as well as space (see for example Stringer 1988; Grün & Stringer 1991; Bar-Yosef 1992; Rak 1993). However, other workers have argued for a return to the views of McCown & Keith, for only a single morphologically variable population during the Middle Paleolithic. For example, Wolpoff (1992) has stated "the amount of variation in measurements from the Middle Paleolithic people from the Levant appears to be less than that in a modern population." This view was challenged by van Vark & Bilsborough (1991), who used Howells' modern data base to show that the generalized distance (D^2) between an Israeli Neanderthal specimen (Tabun) and three early modern specimens (Skhul 5, Qafzeh 6 and 9) was significantly greater than the distance between recent "Eskimo" and "Bushman" samples, supporting Rak's view (as cited in Culotta 1991). Wolpoff (1992) used ranges of 13 cranial measurements for the Israeli fossils compared with those of a larger British sample to show that the latter single "population" had the greater degree of variation, but he did not address the issue of differences in pattern. When, instead, the variation in measurements is compared using the respective standard deviations, the Levantine sample is certainly more variable overall than the much larger recent one (van Vark, pers. comm.). Wolpoff's (1992) comparisons of the variation in Middle Paleolithic and recent samples have also been strongly criticized on statistical grounds by Foote (1993).

I think it is the pattern of measurement differences which seem most important here, and it has been demonstrated many times that the Levantine Neanderthals (as well as those from Shanidar) are distinct metrically and morphologically from the Skhul-Qafzeh sample (Stringer 1974, 1978, 1989, 1992b; Stringer & Trinkaus 1981; Vandermeersch 1981; Trinkaus 1983, 1984, 1992). The Asian Neanderthals show the presence of Neanderthal apomorphies of the face, mandible, cranial vault, and body shape, but some of these features occur at lower frequencies than is the case for the European Neanderthals. Thus for the standardized measurements of Table 2, the Asian Neanderthals are like the European Neanderthals in four measurements (NLH, SSS, FRS, and OCF; see also Stringer 1992b), but are more like archaic African, and modern, crania in two others (BNL, PAS). A whole suite of postcranial features also distinguish the Asian Neanderthals from the Skhul-Qafzeh samples, although many of these are likely to be plesiomorphous in the Neanderthals, including those of the pubic ramus, as already discussed (Trinkaus 1983, 1984, 1986, 1992; Rak 1990, 1993). There are also a few similarities between some of the Asian Neanderthals and the Skhul-Qafzeh samples. These include the flatter midfaces of Tabun 1 and

Shanidar 2 and 4, and the parietal proportions of Shanidar 1. The former features may be related to the greater age (and/or plesiomorphous nature) of the specimens or to regional variation in the Neanderthals, while the latter could represent an individual variation or a reflection of gene flow from non-Neanderthals. Phenetic comparisons (Stringer 1991) suggest that the Saccopastore early Neanderthals, the late European, and the Asian Neanderthals, are about equally related to each other, and it is possible that some of the early Neanderthal morphology (for example rounder occipital profile, larger mastoid processes) was retained to a greater extent in the Asian Neanderthals. On the other hand, without yet having clear data on the body shape of the early Neanderthals, we are presently unable to determine whether the presence of similar limb proportions in the European and Asian Neanderthals is due to common inheritance, continuing gene flow, or (less probably) homoplasy.

Thus, the kind of systematic approach that I (and others like Vandermeersch, Trinkaus, and Rak) use, recognizes a significant division between the Asian Neanderthals and the Skhul-Qafzeh sample, whether this is regarded as subspecific or specific. This approach is disparagingly referred to as "replacement systematics" by Clark & Lindly (1989), but I would prefer to use another r-word — "reality systematics" — for this division. Two "morphs" are recognizable, and while they overlap in some respects they are quite distinct in others. One is Neanderthal-like in predominant pattern, while the other has been termed "anatomically modern" by a number of workers.

A clear initial impression of the degree of modern affinity can be gained by a comparison of the fossil group means with the overall means for modern humans obtained from Howells' large (n = *circa* 2,400) recent cranial series. The early Upper Paleolithic sample is distinct at the ± 2 standard deviation level in only one out of 36 measurements (skull length), the Skhul-Qafzeh group in 6 measurements, the African late archaics in 12, the Asian Neanderthals in 16, and the European Neanderthals in 20. While such a simple comparison does not distinguish factors of shape from size, there is evidently a gulf in modern affinities here between Skhul-Qafzeh and the two Neanderthal samples, and this brings us on to Corruccini's (1992) views on the Levantine fossils (Proposal 4).

In an interesting review of the Skhul material, Corruccini stated that Skhul 5 was usually the only specimen considered in discussion of the modernity of the material, and that when Skhul 4 and 9 were considered as well, metrical variation produced a much greater overlap with Neanderthals. He also claimed that Qafzeh 6 was cranio-phenetically closer to Neanderthals than to a European Upper Paleolithic sample. Dealing with the first point, I have always used Skhul 9 in comparisons where it provided relevant data (for example Stringer 1974, 1978, 1989, 1992b; Stringer & Trinkaus 1981), and as a specimen under my care, it is very familiar to me. However it is, *contra* Corruccini, far less complete and well-preserved than Skhul 5, and therefore of more limited value. I would agree that its parietal region is more primitive in shape than that of Skhul 5 (see for example Stringer 1978; Stringer & Trinkaus 1981), but I would argue that this is not necessarily a Neanderthal feature, and could relate it just as easily to late archaic African specimens (see the comparisons of Table

2 for parietal subtense, PAS). When the pelvis and femur of Skhul 9 are also considered, this specimen shows a high femoral neck angle, rather than the lower angle more characteristic of archaic hominids (Trinkaus 1992). As Corruccini (1992) notes, it does have a relatively long (although quite thick) pubic ramus, but this feature is less Neanderthal-like when standardized for body size (Trinkaus 1984). For Skhul 4, the cranial preservation is arguably worse than for Skhul 9, and I could take virtually no reliable measurements on the specimen, but the postcranial morphology is better preserved and clearly "modern." Unfortunately, in his analyses, Corruccini appears to have used the McCown & Keith (1939) data on Skhul 4 and 9 without any detailed examination of the fossils themselves. Thus he seems to have failed to recognize why most of the cranial measurements published by McCown & Keith are bracketed (they are uncertain or estimated). In McCown & Keith's Tables 73-79, the following number of cranial measurements are given (bracketed number gives numbers of measurements estimated): Skhul 5, 63 (4 estimated); Skhul 4, 61 (37 estimated); and Skhul 9, 54 (41 estimated). Thus Skhul 5 rightfully predominates in discussion about the site because of its better cranial preservation, not because of any inherent bias on the part of researchers.

When we consider the Skhul-Qafzeh sample as a whole, I agree with Corruccini about its phenetically intermediate nature between archaic (not specifically Neanderthal) and recent humans, but I would see this as reflecting an early modern morphology which still retains a number of primitive features. If the Skhul-Qafzeh sample is seen as metrically intermediate between recent and archaic humans, the archaic humans it most clearly resembles are the African late archaics followed (with less certain data) by the Dali-Maba sample (Stringer 1992b, 1994). In attempting to stress the differentiation of the Skhul-Qafzeh group from the Neanderthals, workers such as myself have probably overemphasized the cranial modernity of the specimens (as distinct from their clearer postcranial modernity). However, when the largest possible data sets are used, the Skhul-Qafzeh sample can justifiably be grouped with recent *Homo sapiens* on the basis of synapomorphies such as a rounder cranial profile (frontal, parietal, and occipital shape), mental eminence, pelvic and femoral structure, and symplesiomorphies such as a low face and nasal height combined with a flatter midface. Plesiomorphous retentions such as the broad (but short) face, relatively flatter parietals in some specimens, greater total facial prognathism, broad nose, larger dentition, and palatal dimensions are those which lead to an intermediate phenetic position between archaic and modern, but there is little indication of any specific Neanderthal relationship.

Instead there is evidence of a prior archaic ancestry, which should surprise no one who believes in human evolution — we should not expect 100,000-year-old members of our species to look exactly like living people. Corruccini suggests a revival of the pre-Neanderthal formulation of modern human origins (see for example Howell 1957), and I came close to this conclusion in my earlier research (Stringer 1974). Now I would argue that specimens like Jebel Irhoud are the persisting African equivalents of the pre-Neanderthals — representatives of populations which have retained a rather generalized cranial form from the hypothesized common ancestor of Neanderthals and modern humans. We differ over the data and its interpretations, but both Corruccini and I recognize the intermediate nature of the Skhul-Qafzeh group, although we might place them on opposite sides of any archaic-modern divide. We also recognize the late evolution of the fully modern form and modern regionality, and the probability that the late Neanderthals of Europe were not ancestral to modern humans.

Finally, I agree with Corruccini that the large internal variation of the Skhul-Qafzeh sample is an important and still largely unexplored issue. For example, Qafzeh 6 and 9 differ markedly in facial shape (from my own observations and data, and van Vark & Schaafsma 1992), and there is the possibility that these human samples cover a greater range in time than recently assumed (McDermott *et al.* 1993), which is now under further investigation. Nevertheless, the overall cranial and postcranial data for Qafzeh and Skhul show that in the early Late Pleistocene Israel was inhabited by people showing clear synapomorphies with modern humans, while Neanderthals persisted in western Eurasia and while the Ngandong people may still have lived in Java. Africa is the only other region with comparably early evidence of modern human apomorphies, even though the evidence is either less complete or less well dated (for example the Klasies fronto-nasal fragment 16425, mandible 16424, Omo Kibish 1, Mumba, Die Kelders, Guomde KNM-ER 999 and 3884, and, possibly, Border Cave 1, 2, 3, and 5; see for example Bräuer 1992; Trinkaus 1993).

Concluding Comments

Some workers have evidently become tired of discussions about hominid systematics and seem to wish to set aside the whole process of classifying fossil material (*e.g.*, Trinkaus, in press). However, this process is fundamental to the proper study of the material, whether we are carrying out comparative functional analyses, reconstructing phylogeny, or creating evolutionary scenarios. All of these require the comparison of fossil units (morphs), and the construction of, and relationship between, these morphs cannot just be set aside, whatever difficulties may be entailed. I believe that the most distinct Middle-Late Pleistocene morphs should be allocated to distinct species. Pleistocene *Homo* (excluding *Homo habilis* from consideration) is therefore most reasonably divided into a minimum of three species: *erectus, neanderthalensis*, and *sapiens*. At least one further species should be recognized, which probably represents the stem species from which both *Homo neanderthalensis* and *Homo sapiens* evolved, assuming that they did not descend directly out of *Homo erectus*. This species may be named *Homo heidelbergensis*, *Homo rhodesiensis*, or *Homo helmei*, and it will require further study of existing fossils, as well as new discoveries, to resolve this very current issue satisfactorily.

Acknowledgements

I would first like to thank the organizers and sponsors of the meeting for inviting and supporting my participation in an excellent Symposium. I would also, as always, like to thank all those who have given me study access to fossil hominid material.

Finally I would like to thank Robert Kruszynski for his help in preparing the figures, and Irene Baxter for her help in preparing the manuscript.

Literature Cited

Andrews, P. 1984. An alternative interpretation of the characters used to define *Homo erectus*. *Cour. Forschungsinst. Senckenb.* 69:167-175.

Arsuaga, J.-L., I. Martinez, A. Gracia, J.-M. Carretero, & E. Carbonell. 1993. Three new human skulls from the Sima de los Huesos Middle Pleistocene site in Sierra de Atapuerca, Spain. *Nature* 362:534-537.

Bar-Yosef, O. 1992. The role of western Asia in modern human origins. *Philos. Trans. R. Soc. London* B 337:193-200.

Bräuer, G. 1992. Africa's place in the evolution of *Homo sapiens*. Pages 83-98 *in* G. Bräuer & F. H. Smith, eds., *Continuity or Replacement: Controversies in* Homo sapiens *Evolution*. Balkema, Rotterdam, The Netherlands.

Clark, G. A., & J. M. Lindly. 1989. The case for continuity: Observations on the biocultural transition in Europe and Western Asia. Pages 626-676 *in* P. Mellars & C. Stringer, eds., *The Human Revolution*. Edinburgh University Press, Edinburgh, UK.

Clarke, R. J. 1990. The Ndutu cranium and the origin of *Homo sapiens*. *Jour. Hum. Evol.* 19:699-736.

Corruccini, R. S. 1987. Shape in morphometrics: Comparative analyses. *American Jour. Phys. Anthropol.* 73:289-303.

_____. 1992. Metrical reconsideration of the Skhul IV and IX and Border Cave 1 crania in the context of modern human origins. *American Jour. Phys. Anthropol.* 87:433-446.

Culotta, E. 1991. Pulling Neandertals back into our family tree. *Science* 252:376.

Dreyer, T. F. 1935. A human skull from Florisbad, Orange Free State, with a note on the endocranial cast, by C. U. Ariëns Kappers. *Proc. K. Ned. Akad. Wet.* 38:119-128.

Foote, M. 1993. Human cranial variability: A methodological comment. *American Jour. Phys. Anthropol.* 90:377-379.

Frayer, D. W. 1992. The persistence of Neanderthal features in post-Neanderthal Europeans. Pages 179-188 *in* G. Bräuer & F. H. Smith, eds., *Continuity or Replacement: Controversies in* Homo sapiens *Evolution*. Balkema, Rotterdam, The Netherlands.

Groves, C. P. 1989. A regional approach to the problem of the origin of modern humans in Australasia. Pages 274-285 *in* P. Mellars & C. Stringer, eds., *The Human Revolution*. Edinburgh University Press, Edinburgh, UK.

Grün, R., & C. Stringer. 1991. Electron spin resonance dating and the evolution of modern humans. *Archaeometry* 33:153-199.

Harrison, T. 1993. Cladistic concepts and the species problem in hominoid evolution. Pages 345-372 *in* W. H. Kimbel & L. B. Martin, eds., *Species, Species Concepts, and Primate Evolution*. Plenum Press, New York.

Howell, F. C. 1957. The evolutionary significance of variation and varieties of "Neanderthal" man. *Quart. Rev. Biol.* 32:330-347.

Howells, W. W. 1973. *Cranial Variation in Man: A Study by Multivariate Analysis of Patterns of Differences Among Recent Human Populations*. Pap. Peabody Mus. Archaeol. and Ethnol. Harvard Univ. 67.

_____. 1989. *Skull Shapes and the Map: Craniometric Analyses in the Dispersion of Modern Homo*. Pap. Peabody Mus. Archaeol. Ethnol. Harvard Univ. 79.

Kimbel, W. H., & L. B. Martin, eds. 1993. *Species, Species Concepts, and Primate Evolution*. Plenum Press, New York. 560 pp.

Kimbel, W. H., & Y. Rak. 1993. The importance of species taxa in paleoanthropology and an argument for the phylogenetic concept of the species category. Pages 461-485 *in* W. H. Kimbel & L. B. Martin, eds., *Species, Species Concepts, and Primate Evolution*. Plenum Press, New York.

King, W. 1864. The reputed fossil man of the Neanderthal. *Quart. Jour. Sci.* 1:88-97.

Lahr, M. M. 1994. The multiregional model of modern human origins: A reassessment of its morphological basis. *Jour. Hum. Evol.* 26:23-56.

Li, T., & D. Etler. 1992. New Middle Pleistocene crania from Yunxian in China. *Nature* 357:404-407.

McCown, T. D., & A. Keith. 1939. *The Stone Age of Mount Carmel II: The Fossil Human Remains from the Levalloiso-Mousterian*. Clarendon Press, Oxford, UK. 390 pp.

McDermott, F., R. Grün, C. B. Stringer, & C. J. Hawkesworth. 1993. Mass spectrometric U-series dates for Israeli Neanderthal/early modern hominid sites. *Nature* 363:252-255.

Pope, G. G. 1992. Craniofacial evidence for the origin of modern humans in China. *Yearb. Phys. Anthropol.* 35:243-298.

Rak, Y. 1990. On the differences between two pelvises of Mousterian context from the Qafzeh and Kebara caves, Israel. *American Jour. Phys. Anthropol.* 81:323-332.

_____. 1993. Morphological variation in *Homo neanderthalensis* and *Homo sapiens* in the Levant: A biogeographic model. Pages 523-536 *in* W. H. Kimbel & L. B. Martin, eds., *Species, Species Concepts, and Primate Evolution*. Plenum Press, New York.

Rightmire, G. P. 1990. *The Evolution of* Homo erectus. Cambridge University Press, Cambridge, UK. 260 pp.

Ruff, C. 1991. Climate and body shape in hominid evolution. *Jour. Hum. Evol.* 21:81-105.

Smith, F. H., & E. Trinkaus. 1991. Les origines de l'homme moderne en Europe centrale: un cas de continuité. Pages 251-290 *in* J.-J. Hublin & M.-A. Tillier, eds., *Aux origines d'* Homo sapiens. Presses Universitaires de France, Paris, France.

Stringer, C. B. 1974. Population relationships of later Pleistocene hominids: A multivariate study of available crania. *Jour. Archaeol. Sci.* 6:235-253.

_____. 1978. Some problems in Middle and Upper Pleistocene hominid relationships. Pages 395-418 *in* D. Chivers & K. Joysey, eds., *Recent Advances in Primatology 3: Evolution*. Academic Press, London, UK.

_____. 1981. The dating of European Middle Pleistocene hominids and the existence of *Homo erectus* in Europe. *Anthropologie* (Brno) 19:3-14.

_____. 1984. The definition of *Homo erectus* and the existence of the species in Africa and Europe. *Cour. Forschungsinst. Senckenb.* 69:131-144.

_____. 1985. Middle Pleistocene hominid variability and the origin of Late Pleistocene humans. Pages 289-295 *in* E. Delson, ed., *Ancestors: The Hard Evidence*. Alan R. Liss, New York.

_____. 1988. The dates of Eden. *Nature* 331:565-6.

_____. 1989. The origin of early modern humans: A comparison of the European and non-European evidence. Pages 232-244 *in* P. Mellars & C. Stringer, eds., *The Human Revolution*. Edinburgh University Press, Edinburgh, UK.

_____. 1991. A metrical study of the Guattari and Saccopastore crania. *Quaternaria Nova* 1:621-638.

_____. 1992a. Replacement, continuity and the origin of *Homo sapiens*. Pages 9-24 *in* G. Bräuer & F. H. Smith, eds., *Continuity or Replacement: Controversies in* Homo sapiens *Evolution*. Balkema, Rotterdam, The Netherlands.

_____. 1992b. Reconstructing recent human evolution. *Philos. Trans. R. Soc. London* B 337:217-224.

_____. 1992c. The evolution of *Homo sapiens*. Pages 43-94 *in* M. J. Adler, ed., *The Great Ideas Today*. Encyclopedia Britannica, Chicago, Illinois.

_____. 1993. Secrets of the pit of the bones. *Nature* 362:501-502.

_____. 1994. Out of Africa: A personal history. Pages 149-172 *in* M. H. Nitecki & D. V. Nitecki, eds., *Origin of Anatomically Modern Humans*. Plenum Press, New York.

_____, & G. Bräuer. 1994. Methods, misreading, and bias. *American Anthropol.* 96:416-424.

_____, & E. Trinkaus. 1981. The Shanidar Neanderthal crania. Pages 129-165 *in* C. B. Stringer, ed., *Aspects of Human Evolution*. Taylor & Francis, London, UK.

Tattersall, I. 1986. Species recognition in human paleontology. *Jour. Hum. Evol.* 15:165-175.

_____. 1992. Species concepts and species identification in human evolution. *Jour. Hum. Evol.* 22:341-349.

Thorne, A. G., & M. H. Wolpoff. 1992. All about Eve. *Sci. American* 267(3):6.

Trinkaus, E. 1981. Neanderthal limb proportions and cold adaptation. Pages 187-224 *in* C. B. Stringer, ed., *Aspects of Human Evolution*. Taylor & Francis, London, UK.

_____. 1983. *The Shanidar Neandertals*. Academic Press, New York. 502 pp.

_____. 1984. Western Asia. Pages 251-325 *in* F. H. Smith & F. Spencer, eds., *The Origins of Modern Humans*. Alan R. Liss, New York.

_____. 1986. The Neandertals and modern human origins. *Annu. Rev. Anthropol.* 15:193-218.

_____. 1991. The evolution and dispersal of modern humans in Asia. *Curr. Anthropol.* 32:353-355.

_____. 1992. Morphological contrasts between the Near Eastern Qafzeh-Skhul and late archaic samples: grounds for a behavioral difference? Pages 277-294 *in* T. Akazawa, K. Aoki, & T. Kimura, eds., *The Evolution and Dispersal of Modern Humans in Asia. Hokusen-Sha, Tokyo, Japan.*

_____. 1993. A note on the KNM-ER 999 hominid femur. *Jour. Hum. Evol.* 24:493-504.

_____. (in press). Brains and bodies: mosaic trends in Middle Pleistocene archaic *Homo* morphology. *In* J. M. Bermudez de Castro, ed., *Human Evolution in Europe and the Atapuerca Evidence*. Museo Nacional de Ciencias Naturales, Madrid, Spain.

Vandermeersch, B. 1981. *Les Hommes Fossiles de Qafzeh (Israel)*. Centre National du Recherche Scientifique, Paris, France. 319 pp.

van Vark, G. N., & A. Bilsborough. 1991. Shaking the family tree. *Science* 253:834.

_____. & W. Schaafsma. 1992. Advances in the quantitative analysis of skeletal morphology. Pages 229-261 *in* S. R. Saunders & M. A. Katzenberg, eds., *Skeletal Biology of Past Peoples: Advances in Research Methods*. Wiley, New York.

Wolpoff, M. H. 1980. Cranial remains of Middle Pleistocene European hominids. *Jour. Hum. Evol.* 9:339-358.

_____. 1989. Multiregional evolution: The fossil alternative to Eden. Pages 62-108 *in* P. Mellars & C. B. Stringer, eds., *The Human Revolution*. Edinburgh University Press, Edinburgh, UK.

_____. 1992. Levantines & Londoners. *Science* 255:142.

_____, X. Wu, & A. G. Thorne. 1984. Modern *Homo sapiens* origins: A general theory of hominid evolution involving the fossil evidence from East Asia. Pages 411-483 *in* F. H. Smith & F. Spencer, eds., *The Origins of Modern Humans*. Alan R. Liss, New York.

_____, A. G. Thorne, J. Jelínek, & Y. Zhang. 1994. The case for sinking *Homo erectus*: 100 years of *Pithecanthropus* is enough! *Cour. Forschungsinst. Senckenb.* 171:341-361.

Wood, B. 1992. Origin and evolution of the genus *Homo*. *Nature* 355:783-790.

Wright, R. S. V. 1992. Correlation between cranial form and geography in *Homo sapiens*:

CRANID — A computer program for forensic and other applications. *Archaeol. Oceania* 27:105-112.

Behavior and Human Evolution

Alison S. Brooks
Department of Anthropology
George Washington University
Washington, DC 20057

Behavior is critical to an understanding of human evolution, as it underlies not only definitions of "humanness" but also some of the major species distinctions within the genus *Homo*. Yet behavioral and morphological shifts in the evolutionary history of *Homo* do not always coincide. Fossil behavior can be reconstructed both from the morphology, chemistry, pathology, and disposition of the fossils themselves and from the artifacts, faunal remains, and sites of the archaeological record. A review of behavioral evolution in the genus *Homo* suggests that we do not yet have behavioral evidence to support more than one species of earliest *Homo* between 2.3 and 1.6 Ma. Niche differentiation in behavior may exist, but is either not reflected in the record or not accessed by our current analysis techniques. Similarly, although the replacement of *H. habilis* by *H. erectus* (*sensu lato*) is indeed accompanied by major behavioral change, the later shift from *H. erectus* to *H. sapiens* is less clearly associated with behavioral evolution. Finally, there is evidence for a major behavioral change coinciding with the appearance of anatomically modern humans, *H. sapiens sapiens*, which includes the earliest appearance of this taxon in Africa. In this case, the behavioral evidence supports separation at the species level of modern from archaic *sapiens*.

Arguably the most frequently asked question in public symposia on human evolution is: "How human are the fossil taxa variously assigned to *Australopithecus* or *Homo*?" What is humanness anyway? Few would argue that the essence of humanness is walking on two legs, although that is how we used to define "hominid" before genetics, paleontology, and cladistic methodology suggested that this term might cover the African apes as well (Delson *et al.* 1977; Schwartz *et al.* 1978). Humanness is more likely to be thought of in behavioral terms: technological, social, cognitive, aesthetic, moral, or symbolic behavior.

How do we know the thoughts of early hominids, or what they did on a daily basis? Did they live in a world that was partly created or built by themselves through the making of tools, buildings, and public and private spaces? Did they know where their next meal was coming from, or just run into it in a haphazard sort of way? Did they mate for the long term and maintain ties to adult children throughout their lives? Did they achieve senior citizen status, care for each other's wounds and illnesses, or bury their dead? Did they enjoy the company of friends from afar or decorate themselves

Contemporary Issues in Human Evolution
Editors, W.E. Meikle, F.C. Howell, & N.G. Jablonski

and their environment? Was meaning invested in arbitrary symbols: sounds, shapes, colors, or gestures?

In biology, not only is behavior an important component of an animal's ecology, but often part of what separates it as a species from closely related forms (Dobzhansky 1937; Mayr 1963). Behavior can keep members of different species from meeting, mating, or raising successful offspring. In other words, if a potential mate's dance does not turn you on, you might as well be living on different continents. Behavior is thus integral to an understanding of the species definition, although both systematists and paleontologists try to base the actual definitions entirely on morphological data (Eldredge 1986, 1993; Tattersall 1986; Albrecht & Miller 1993; Kimbel & Martin 1993; Kimbel & Rak 1993). Darwin (1859:76) and his successors (Dobzhansky 1937; Mayr 1963) have generally argued that closely related species experience the greatest competition for resources and can only coexist if they use the environment differently. By analogy, phylogenetic species in the same environment should also differ in their niche behavior (Mayr 1950). An absence of behavioral differentiation would call the species distinction itself into question.

In the study of human evolution, species definitions have often been underlain, overtly or covertly, by supposed behavioral distinctions. The clearest example is *Homo habilis*, a taxon established initially to distinguish the first tool-making forms from non-tool-making predecessors,[1] although some of the eventually included specimens (*e.g.*, OH 24) overlapped to a marked degree morphologically with *Australopithecus*, while some later forms exhibited certain convergences with fossils previously assigned to *Homo erectus*. In fact, the inclusion of *habilis* within *Homo* required a change in then-prevailing morphological definitions to accommodate the smaller cranial capacities typical of the *habilis* group (Leakey *et al.* 1964; Napier & Tobias 1964).

The evolutionary history of our own genus, *Homo*, usually involves the recognition of at least three different species, *H. habilis*, *H. erectus*, and *H. sapiens*, and at least two subspecies of the latter: archaic *Homo sapiens* (and/or *Homo sapiens neanderthalensis*) and *Homo sapiens sapiens*, anatomically modern humans.[2] The morphological boundaries are defined to a great extent by reference firstly to increasing brain size with its accompanying changes in cranial shape, and secondly to decreasing dental size with the accompanying reduction in the face and cranial structures associated with mastication (McHenry 1982). Are there behavioral differences between these forms as well? How can we determine their behavior? The remainder of this paper will examine the behavioral evolution of *Homo* and discuss the fit between the evolution of humanness and the evolution of *Homo*, as defined morphologically.

The Study of Behavior

Sources of information on behavior are limited to the evidence from the fossils themselves and the evidence from the associated artifacts, animal bones, and landscapes which can be termed the archaeological record. A modern analogy, observed or experimental, also underlies most behavioral inferences. We compare the evidence from the fossil or archaeological records with natural bone accumulations (Tappen 1992), the wear on modern teeth (Brace *et al.* 1981) or stone tools (Odell 1977; Keeley 1980; Vaughan 1985), tooth marks on bones in zoo cages (Haynes 1981) or hyena dens (Sutcliffe 1970; Avery 1984; Potts *et al.* 1988), effectiveness of projectile shape determined by experiment (Knecht 1993), relationship of sexual dimorphism to mating behavior in modern apes (see review by Frayer & Wolpoff 1985), or of landscape aridity to group organization in modern hunter-gatherers (Binford 1980, 1982, 1983; Laden & Brooks 1994; Brooks & Laden, in press). These always carry a caution that modern analogues or experiments may not adequately represent the fossil conditions, environments, or behaviors.

Behavior may be inferred from fossils in a number of ways, including their morphology (and signs of violence, stress, or disease), their chemistry, and the disposition of the fossils themselves, in a hyena den or in a cemetery, as an articulated skeleton or a jumble of skulls bearing cutmarks (Table 1).

In inferring behavior from morphology, one might use limb proportions and the shape of the digits to determine the habitual ecological niche. Did the species still have short legs, long arms, and curved digits for sleeping in the trees (Susman *et al.* 1984)? Are there adaptations for heat loss, such as bipedalism itself (Wheeler 1991a, 1991b, 1992a, 1992b), or aspects of the venous flow to and from the brain (Falk 1986, 1990; but see Wheeler 1990; Dean 1990), which might imply activity in the heat of

TABLE 1. Behavior from Fossils

Morphology
Habitual ecological niche: trees/ground, diurnal/nocturnal
Locomotor pattern
Manual dexterity—need to grasp branches or manipulate
Need for learning, memory storage, pattern recognition
Language capability
Diet—coarseness, quality (calories/gm.)
Diet—use of tools to pre-crush or predigest food
Lifespan—structure and length
  Infancy
  Adolescence
Male-female relationships
Male-male relationships
Disease and stress, inter- and intra-group relationships

Chemistry
Diet—quality, meat vs. vegetable food, roots vs. leaves
Ecological niche
Climate

Post-mortem treatment
Consumption or predation
Primary burial, with or without artifacts/grave goods
Secondary burial, defleshing
Cemeteries—have settlement pattern implications

the day to avoid predators? How efficient was their bipedalism (Taylor & Rowntree 1973; Rodman & McHenry 1980; Rak 1991)? The curvature of the thumb, its joint with the wrist bones, and shape of its tip might imply capacity for fine manipulation of objects (Susman 1988; Ricklan 1987; Tobias 1991). Hyper-robust limbs, joints, and muscle markings imply extreme physical stress in hunting or defense (Trinkaus 1983b, 1986) or development of a calcium reserve in a shift to carnivory (Kennedy 1992), while more gracile limbs may imply long-distance or tool-mediated hunting (Brace 1992, 1995). The size of the brain, and certain aspects of its shape such as asymmetry, arrangement of sulci, or presence of areas in the left temporal lobe, imply selection for cognitive skills: learning, memory storage, and pattern recognition, as well as for symbolic communication or language (Tobias 1991; Falk 1987a, 1987b). The shape of the cranial base and the vocal tract may also suggest ability to speak (Laitman *et al.* 1979; Lieberman 1984). Body size or stature is a clue to dietary shifts (Ruff 1987; Ruff & Walker 1993). Primates in open savannas are generally baboon-size or smaller; australopithecines are similar, but in *Homo erectus* we see a major size increase implying a shift in dietary quality (calories/g). Poor quality foods are not easily transported and shared owing to their bulk, so food quality might even imply social organization or lack thereof (Foley 1987; Tooby & DeVore 1987, *contra* Lovejoy 1981). Body size and shape may also relate to climate or activity levels (Ruff *et al.* 1993, 1994).

Teeth are a major source of information about not only diet but also technology and social life (see review by Gordon 1993). Smaller chewing teeth on a bigger animal imply either a major dietary shift to softer foods or preparation and predigestion of those foods outside the mouth by pounding or cooking. Smaller and/or monomorphic canines may be a secondary result of post-canine megadonty, may imply monoga-mous pair-bonding and territoriality, as among the gibbons, or even the invention of new and different ways for males to compete for females. Enamel thickness may reflect the hardness or grittiness of food sources (Martin 1985), while enamel structure and tooth eruption patterns tell us about how rapidly the teeth form, and thus how rapidly an infant grows up (Smith 1991). Finally, the wear on teeth can indicate use as tools for manufacturing, stripping, or clamping (Trinkaus 1983a) in the anterior part of the mouth where chewing stress is unlikely, as well as how early in life babies begin to eat solid food (Skinner 1981, 1989). This last is not trivial, since our present population crisis may be partially attributed to earlier weaning of the young. Are males much bigger than females, implying possible extreme competition for mates and harem-formation, as in gorillas? (See, for example, reviews of this issue for *A. afarensis* in McHenry 1991 and Kimbel *et al.* 1994.)

A shift to more predatory food procurement strategies was not without cost (Shipman & Walker 1989). Signs of stress such as enamel hypoplasias and Harris' lines tell us how reliable the food supply was, while injuries attest to predation, as in the case of an *Australopithecus* with leopard tooth marks in its skull (Brain 1968, 1981), and interpersonal violence (*e.g.*, the inflamed knife wound in the back (ninth rib) of a Shanidar Neandertal: Trinkaus 1983a). Healed injuries, such as those of the

Shanidar Neandertal with a healed severe injury to the right arm, may also point to the emergence of moral behavior.

The relatively new field of bone biochemistry can indicate both diet and habitat of the past. The concentration of strontium (Schoeninger 1979; Sillen 1981, 1992, 1993) tells us how much meat (or roots) a fossil human ate, while the concentration of carbon-13 tells us how much of the diet was derived from tropical grasses or aquatic resources (Price *et al.* 1985; Tuross & Fogel 1993). Nitrogen isotopes can indicate dietary quality (Schoeninger *et al.* 1983), weaning age (Fogel *et al.* 1989), or whether the individual was starving at the time of its death (Tuross & Fogel 1993). Both carbon and oxygen isotopes can also indicate climate: oxygen-18 is more concentrated in colder periods, while tropical grasses with lower ^{13}C values will expand in dry periods at the expense of forests (Cerling 1992).

Finally, the disposition of the remains is a primary indication of behavior. Are these fossils lying under their living sites or in the lairs of leopards (Brain 1981)? Are they buried deliberately or simply left out haphazardly? Are the bones articulated, or are there signs that this is a collection of disjointed parts in a secondary burial? Are there cutmarks on the bones? Were these made when the bones were fresh, suggesting cannibalism or scalping (Smith 1976; Trinkaus 1985; White 1986), or when the bones were dried, during defleshing and secondary burial? Are a number of individuals buried together in a cemetery or mass grave such as the epipaleolithic one near Jebel Sahaba, Egypt (Wendorf 1968), or the large concentration of "pre-Neandertals" from Atapuerca in Spain (Bahn 1996; Arsuaga *et al.* 1993)? Is this a mass disaster, indicated by a single set of prehistoric excavations or structures, or a sequential use of the place for this special purpose over a long period of time, indicated by overlapping sediment accumulation pits, or structures?

A somewhat different range of behaviors may be derived from examining the archaeological record (Table 2). Stone tools can be analyzed in terms of raw material, technology of manufacture, function, and style or design, as well as how each of these changes over time and space. The distance the raw material was carried from its source tells us about cognitive planning for aesthetic and function-specific ends, about locomotor behavior, ranging pattern, and even intergroup relations if trade was involved (Toth & Schick 1993; Merrick *et al.* 1994). The process of manufacture may indicate cognitive organization and handedness (Toth 1987) or the ability of the artisan to visualize the final shape in the mind, as well as the technological sophistication and understanding of material properties of particular stones, much as Michelangelo envisaged the "David" in a block of damaged marble. What the tool was used for can be determined from patterns of damage and from residues (Loy 1983) or polishes (Keeley 1980) by reference to experimentally-induced wear and damage. What happens to a bone knife if you cut up a mammoth with it? At the Smithsonian, several years ago, my colleagues spent a few days cutting up a dead elephant named Ginsburg in order to find out (Stanford 1979). The shapes of tools, and the degree to which size and shape were standardized, can indicate hafting or use of projectiles (Shea 1988), both of which require more standardization. Alternatively, standardization may simply suggest the cognitive skills of the makers. If shape

TABLE 2. Behavior from Archaeology

Stones
Raw material source — transport and trade
Shape: cognitive imagination, thinking ahead
Functions — (from residues or wear)
Use of symbols — in shape or design
Rates of change in space and time, similarity to "culture"
Re-use and reworking, planning, landscape use, raw material conservation

Bones
Species represented: large or small, easy or dangerous, herd or individual
Ages and sexes: human hunting competence, methods, selectivity
Skeletal parts: scavenging or hunting, transport or on-the-spot consumption,
    surfeit or scarcity
Season of death: seasonality and scheduling of landscape use
Damage, breakage, cutmarks and toothmarks: use of animal products,
    scavenging or hunting, site-use pattern

Sites
Distribution of residues on the landscape
Group movements, scheduling and re-use, cognitive maps
Relation to paleogeography, why were certain places important
Relation to paleoenvironment
Comparison of site contents: activity differences in space
Status or class differences
Gender bias in the record

changes dramatically from one region to the next, we may be looking at ethnic or
cultural differences, symbols of group identity (Wiessner 1983; Sackett 1977, 1980,
1982; Wilmsen 1974; Hodder 1985). Finally, the life history of the tool after it was
made, whether it was re-used, re-sharpened, or dropped and later recycled into
something else tells us about re-occupation and scheduling of landscape use (Binford
1979; Kuhn 1991, 1992, 1995). Recycling may indicate higher levels of "humanness," requiring a concept of the future, or simply a greater degree of territoriality
and re-use.

Bones of prey animals also carry behavioral information. Were the species
involved large or small, gregarious or solitary, dangerous or easy to kill (Klein 1978)?
Each of these implies certain needed technological or even social and cooperative
skills to hunt successfully. Ages and sexes of the prey also suggest the competence
of the hunter, since competent humans may tend to take more big prime age males
than other carnivores (Brain 1981; Klein 1982; Kuhn & Stiner 1994; Stiner 1994). If
all ages and sexes are present — a "catastrophic" mortality profile — the hunt may
have been a wholesale slaughter or ambush of an entire group. The skeletal parts of
the prey animal have behavioral implications: lots of limb bones may mean the meat
was transported to the site, while the axial skeleton and skull of a big animal might
indicate the location where the animal was actually butchered or eaten (*e.g.*, Bunn &
Kroll 1986). Was the meat obtained by scavenging or by hunting? Over-repre-
sentation of distal limb segments or skulls may indicate scavenging, what was left

after the carnivores were finished. The presence of articulated skeletons among the prey may show that they had much more than they needed for food (Frison 1993). Seasonal scheduling of visits to the site will result in prey animals who all died in the same season. This can be demonstrated through study of the growth layers in tooth roots (*e.g.*, Lieberman 1993; Spiess 1979). The pattern of damage and breakage of bone and the presence of carnivore toothmarks are major clues to behavior. For example, stone tool cutmarks over toothmarks of carnivores might imply scavenging, while toothmarks over cutmarks imply that the people did not stay in one place long (Blumenschine *et al.* 1994). Carnivores would not visit the site while the people were there, and bones lose interest for most carnivores after a week or two (Yellen 1991).

Archaeological sites, or the distribution of residues on the landscape, are perhaps the most interesting and neglected aspect of archaeology; indeed, the very existence of such concentrations was a major feature of the initial development of *Homo*. In nomadic hunter-gatherer groups, visible residues are most likely to result from re-occupation rather than from a single visit (Brooks & Yellen 1987; Binford 1983). Re-use itself implies a cognitive map of the landscape, and might indicate a seasonal round of activities.

Site locations may imply dietary selection and procurement strategy (Stahl 1984; Sept 1992). What are the landscape features associated with such concentrations? High points? Caves? Open air locations? River margins? Gallery forest? Stone outcrops? How do these change as species of *Homo* change, or with changing environments? Finally, are all sites the same in what they contain? Are some sites full of projectiles, while others are full of scrapers, possibly representing male and female activities respectively (Binford & Binford 1966)? Are some sites richer in exotic raw materials, ornaments, or specially constructed features than others? Could these be ceremonial centers, or Nob Hills of the past where some individuals controlled a disproportionate share of the scarce goods (*e.g.*, Soffer 1985)?

Behavioral Evolution in *Homo*

From the twin perspectives of the archaeological and fossil records, we turn now to a consideration of species definition in *Homo*. What are the major behavioral shifts in human evolution? (See Table 3.) Do they co-vary with major morphological shifts that mark species boundaries? If not, why not? Is it because the two species in question did not differ in behavior, in which case it would be difficult to justify their separation at the species level? Or is it because we cannot document the difference, either because it left no trace or because we are not reading the record correctly?

Homo habilis

Homo habilis, whether *sensu strictu* or *lato* (see Wood, this volume) is generally accorded the primary place at the base of the lineage of *Homo*, beginning as early as 2.3 Ma. It was first defined as a behavioral species, whose morphological consistency was debated from the beginning. In behavioral morphology it is the first non-megadont bipedal hominid, with a significant increase in brain/body-size ratio

TABLE 3. Major Behavioral Shifts in Human Evolution

---

1. Initial concentration of tools and food residues on the landscape, adaptations to heat (*H. habilis*)

2. Habitat and niche expansion to mid-latitude temperate zones, fire and symmetry, projectiles and hunting, long adolescence (*H. erectus*)

3. Habitat and niche expansion to high latitudes, symbolic behavior, long-distance social networks and social complexity, environmental modification (*H. sapiens sapiens*)

(Tobias 1991). The smaller teeth of many individual fossils may indicate the use of tools for pre-processing food. Postcranial morphology and body size, however, may indicate that, like *Australopithecus*, it still spent a lot of time in the trees (Susman *et al.* 1984; Aiello & Wheeler 1995). The thumb's carpo-metacarpal joint and apical tuft or tip are similar to those of the robust australopithecines, implying a similar capability for manipulating tools (Susman 1988; Tobias 1991). The larger brain and possible Broca's area imply more capacity for learning, while the venous drainage pattern may suggest that *H. habilis* was active at midday (Falk 1986, 1987b, 1990, 1992; but see Wheeler 1990; Dean 1990). Although much debated, the enamel growth and eruption pattern of the teeth may suggest that the long-adolescence pattern of humans had not yet developed (Bromage 1987; Smith 1991). Size differences and implied sexual dimorphism are extreme to an extent inconsistent with the common human social pattern of today: pair bonding in a multi-male grouping (Frayer & Wolpoff 1985).

Behavioral archaeology indicates that *H. habilis* was limited to the African tropics and sub-tropics, although a recent much-debated claim places this species in South China as well (Huang *et al.* 1995). Stone tools consist mostly of flakes and cores, with no consistent shapes (Chavaillon 1976; Merrick & Merrick 1976; Kibunjia *et al.* 1992; Toth & Schick 1993; Rogers *et al.* 1994). Raw material was transported about 3 km at Olduvai (Schick & Toth 1993). Sites themselves are important indicators of changing hominid behavior on the landscape, as none occur before 2.5 Ma. They are probably not home bases or living sites, as their locations on lakeshores are not consistent with safe sleeping places (Brooks & Yellen 1987). It is quite probable that hominids were still sleeping in the trees.

Bones imply that most meat was scavenged (Shipman 1986; Potts 1988; Bunn & Kroll 1986). There is no evidence for symbolic behavior. This was not a human way of life. The correspondence between behavior and species boundary is hard to determine, although the first tools at Gona and Lokalele at 2.5 Ma may not significantly predate the first *H. habilis* at 2.3 to 2.1 Ma (Hill *et al.* 1992; Schrenk *et al.* 1993; Rogers *et al.* 1994). *H. habilis*' behavioral niche was not too wide, as this species coexisted throughout with at least one other hominid: *A. robustus/A. boisei*, and possibly with others (see Wood, this volume). Because the two species co-occur, we cannot be sure that *A. boisei/robustus* did not also make tools; their thumb morphology is similar (Susman 1988). One possible clue to the difference between the two is bone biochemistry. *Australopithecus robustus* may have fed on roots and a surprising quantity of meat (Sillen 1993); preliminary data on *Homo*, however, is very similar (Lee-Thorp & Thackery 1996).

Homo erectus

The next species in the lineage of *Homo*, *H. erectus* (or *H. ergaster* if the East African forms are considered a different species as per Wood 1984) appears alongside *H. habilis* and *A. robustus* by 1.7 Ma. This is a major morphological species boundary in human evolution (Rightmire 1990). It used to be considered less significant in terms of behavior, as some "Oldowan" tools continue to be made (Leakey 1971). More recently, however, it is evident that *H. erectus* also represents a major behavioral shift. In behavioral morphology, it had smaller molars but larger body size and brain/body size ratio. This goes against the hypsodont trend of other large mammals at 1.6 Ma, suggesting that *H. erectus* must have solved the food problem in open environments through tools. Enamel microscopy suggests that childhood and adolescence were closer in pattern to ours (Smith 1993; Ruff & Walker 1993). As a long adolescence would prolong the most dangerous period for most mammals, lengthened adolescence itself suggests changes in social life to protect and even feed adolescents. At Nariokotome, a 12-year-old boy was already ca. 5'4-5" tall, implying a modern adult size and large early growth spurt (Walker 1993; Walker & Leakey 1993). In humans this requires an adequate protein source, such as meat provided by hunting. Brain cooling is modern, suggesting diurnal activity in an open environment (Falk 1990, 1992). The presence of asymmetry and possibly of Broca's area imply language capability, although not necessarily speech (Begun & Walker 1993). The limb proportions and curvature of the digits are modern; *H. erectus* did not live in the trees.

The archaeology of *H. erectus* also suggests a major behavioral shift. Its distribution is distinctive; it expands out of the African tropics/subtropics to northwest Africa and the Eurasian subtropics by 1.4 Ma or earlier (Tchernov 1987; Sahnouni 1987; Bar-Yosef 1991; Dzaparidze *et al.* 1992; Swisher *et al.* 1994; Gabunia & Vekua 1995; Huang *et al.* 1995), and well into the temperate zone by 1.1 Ma (Woo 1966; Wu & Olsen 1985; Brooks & Wood 1990; Schick & Dong 1993). This first population explosion implies successful exploitation of a new food source with a wide distribution, as well as the ability to cope with the longer dry seasons or winters of the north. In Africa, *H. erectus* soon outcompeted *Australopithecus*, which was extinct by 1.3 Ma or so.

Tools made by *H. erectus* include handaxes, which are remarkably symmetrical, implying a cognitive preconception of shape, as well as the possible use of projectiles (O'Brien 1981). Handaxes may also have served as curated multi-purpose tools (Toth & Schick 1993; Keeley & Toth 1981) or as well-shaped cores (Potts 1988). Other tools include spheroids, which may have been projectiles or "bolas" (Willoughby 1985) or simply well-used hammerstones (Schick & Toth 1994). Stone transport at Olduvai is now 8 km or more (Toth & Schick 1993). Bone residues imply hunting at some sites such as Olorgesailie (Shipman *et al.* 1981; but see also Binford 1985).[3]

Some bones were heated to very high temperatures at Swartkrans (1.3 Ma) which could indicate cooking or controlled use of fire (Brain & Sillen 1988). At Chesowanja in Kenya, and possibly also at Koobi Fora, burned patches yielding a unique magnetic signal also suggest controlled use of fire by humans at 1.3 Ma (Bellomo 1993, 1994). Sites were located on stream channels and their margins away from lakes (Clark 1970; Toth & Schick 1986, 1993). These could be living sites, although some have many handaxes and spheroids and little else, while others have ordinary flakes. Similarly, a much smaller number of lakeshore sites may contain (Clark *et al.* 1994) or lack (Potts 1994) handaxes. Binford has suggested functional or organizational differences; different tasks may have been carried out at different sites by males vs. females (Binford 1987, 1989), while Potts has argued that the handaxe sites represent evidence for caching stone for later use (Potts 1988).

Caves were occupied in later periods in South Africa (Mason 1988b) and China (Wu & Olsen 1985), implying an ability to hold these successfully, at least intermittently, against other large cave-dwelling mammals, possibly through use of fire (but see Binford & Ho 1985; Binford & Stone 1986; James 1989). Even later *H. erectus* tools in East Asia, however, appear relatively unspecialized (Schick 1994). This is a semi-human way of life.

Homo sapiens

The next species commonly recognized in the *Homo* lineage is *Homo sapiens* "archaic" or *H. heidelbergensis* (see Stringer, this volume). The oldest fossils referred to this group derive from Africa before 500 Ka (Singer & Wymer 1968; Klein & Cruz-Uribe 1991; Clark *et al.* 1994), although some European fossils of possible *sapiens* affinity may be as old (Bahn 1996; Carbonell *et al.* 1995). This is a very poorly defined species boundary in terms of behavior shifts. In behavioral morphology, the brain size increases ca. 33-50%, but the cranial shape, while higher and fuller in the fronto-parietal region, is often broadly similar to that of *H. erectus*. Body size does not increase. Somewhat greater flexion of the cranial base in "archaic" *H. sapiens* may change the shape of the vocal tract, making speech slightly easier (Laitman *et al.* 1979; Lieberman 1984).

Neandertals are often considered as a specialized offshoot of the archaic group: *Homo sapiens neanderthalensis*. Their morphology is characterized by an increase in body mass, joint surfaces, and muscle markings (Trinkaus 1983a, 1983b, 1986), and by a decrease in distal limb segments relative to proximal ones (Holliday & Trinkaus 1991; Ruff 1991). These imply a physical response to ice age scarcity and greater torsional strength of limbs, feet, arms, and hands (Trinkaus *et al.* 1994). Neandertals may have been the first hominids to occupy Europe successfully throughout a pleniglacial, beginning with a cold phase ca. 190 Ka (Tuffreau & Somme 1988). Other peculiarities of Neandertals include possibly a different body stance suggested by the Kebara pelvis (Rak & Arensberg 1987; Rak 1990), a high incidence of Harris' lines and enamel hypoplasias (Ogilvie *et al.* 1989), which suggest stress during growth (Skinner 1989), and little or no wear on the deciduous teeth which could indicate a long nursing period and slow population growth or a more rapid individual transition from infancy to childhood than in ourselves. Neandertals also exhibit maximum sexual dimorphism for *Homo sapiens*. This may imply a different, more harem-like social organization as in the very dimorphic gorillas (Frayer & Wolpoff 1985; Binford 1989) or just greater selection for male effectiveness as providers. Females may also have been more stressed during growth, and thus smaller in stature, than in modern humans. Neither "archaic" *H. sapiens* generally, nor the more specialized Neandertals, exhibit the specific characters of *H. erectus* such as the supraorbital torus (continuous bar of bone over the eyes), sharp angle at the back of the skull (inion), thick cranial bones, and the low cranial vault with maximum breadth in the temporal region. Individually and collectively, however, "archaics" and *H. erectus* fossils represent a continuum with much overlap in particular morphometric traits.

Behavioral archaeology, for early "archaic" *H. sapiens* (*e.g.*, Bodo, Petralona, Ndutu) also suggests little difference from *H. erectus*. Tools are still characterized by handaxe industries at sites such as Ndutu (Mturi 1976; Clarke 1990) and probably also Bodo (Clark *et al.* 1994), and there is little regional differentiation in stone tools. Sites include seasonally-reoccupied open-air sites as well as large mammal butchery sites. Projectiles include wooden spears (Movius 1950). In textbooks, Torralba, Ambrona (elephant butchery), and Terra Amata (open air "huts") are still cited as examples of *H. erectus* behavior, although fossils from nearby sites of comparable age are all "archaic" *H. sapiens* (de Lumley 1969). Hunting may be more competent, as suggested by evidence from the Spanish elephant sites (Villa 1990), but it is hard to compare European sites with preceding African ones on this point due to different availability of both scavengeable carcasses and prey.

The biggest behavioral differences between *H. erectus* and "archaic" *H. sapiens* include the first intensive occupation of most of Europe by the latter (Delson 1991). No undisputed fossils or human occupation sites on that continent are earlier than 600 Ka (Roebroeks 1994), with the possible exceptions of Vallonnet (de Lumley 1988) near Monaco (*ca.* 900 Ka, but with very dubious stone "tools"), Isernia (*ca.* 730 Ka, but the dated tuff may be redeposited from an older context; Peretto 1991), and, at Europe's southeastern edge, a new *Homo erectus* jaw from Dmanisi (Dzaparidze *et al.* 1992; Gabunia & Vekua 1995), in Georgia, possibly dating to 1.4-1.8 Ma.[4] Recently reported human remains from the Sierra de Atapuerca may date to *ca.* 800 Ka, but the hominids already exhibit derived features linking them with later Neandertals (Carbonell *et al.* 1995). *H. sapiens* may not have reached East Asia until significantly later; sites containing remains of *H. erectus* occur in China as recently as 300 Ka (Zhou & Ho 1990).

Additionally, in many "archaic" *H. sapiens* sites the advent of prepared-core technology implies further cognitive imaging of the finished artifact. This technology, however, is also found at several sites in association with late *Homo erectus* (Leakey 1995; McBrearty 1990). A third possible trend is the occupation of more caves and rock-shelters. A greater emphasis on smaller animals in some sites (Kalb *et al.* 1982) may also imply more effective hunting techniques. However, changes in subsistence are very subtle; there is, for example, little exploitation of fish, shellfish, or birds, and few small mammals. The latter may have been substantially removed by carnivore scavenging, however, if hominids moved frequently enough (Yellen 1991). Evidence of symbolic behavior in early "archaics" is limited to possible incisions or engravings on bone (Bordes 1972; Marshack 1989a; but see D'Errico & Villa 1996) and to ocher lumps or "crayons", faceted from use, by 300 Ka (Schmandt-Besserat 1980; Marshack 1989a).

Again, Neandertals are a special case, as they are the only "archaics" who clearly survive well into the time range of *Homo sapiens sapiens* (Lévèque & Vandermeersch 1980; Straus 1994; Straus *et al.* 1993). At least some Neandertals appear to have practiced cannibalism (Le Mort 1989; Defleur *et al.* 1993), while later Neandertals buried their dead with grave goods (Solecki 1971, 1975; Smirnov 1989; Defleur 1993; Hayden 1993; *contra* Gargett 1989). Neandertals used mineral pigments (Bordes 1972), cared for the sick and injured (Trinkaus 1983a), engaged in minimal long-distance trade of shells and stone, and made very occasional ornaments — an amulet with a cross from Tata, a pierced fox canine from La Quina (Marshack 1989a). However, all of these occurrences coincide with or post-date the appearance of these behaviors in connection with modern humans at Qafzeh and Border Cave (Miller *et al.* 1992; Valladas *et al.* 1988; Schwarcz *et al.* 1988). There are more behavioral differences between early "archaics" and Neandertals than between early "archaics" and *H. erectus*, perhaps justifying the recognition by some (*e.g.*, Rak 1993; Stringer 1994) of *H. neanderthalensis* as a separate species of *Homo*. We recognize "humanness" in Neandertals, but it is curiously devoid of the rich symbolic aspects of modern human cultures. Symbols of ethnic identity or of long-distance social networks are absent. Even the linguistic competence of Neandertals is hotly debated, despite the recent recovery of an anatomically "modern" hyoid with the Neandertal from Kebara (Arensberg *et al.* 1990).

Anatomically Modern Humans (*Homo sapiens sapiens*)

The final and current subspecies of *Homo sapiens*, although arguably as distinct morphologically from Neandertals as the latter are from *H. erectus*, is also distinctive in behavioral terms. This distinction is reflected in the literature as "the great leap forward" of Diamond (1992), or the "Upper Paleolithic revolution" of Mellars and others (1989, 1990). In behavioral morphology, the biggest differences include a decrease in robusticity, longer distal limb segments, decrease in tooth size especially in the anterior teeth, facial reduction, and increased basicranial flexion (Brace 1964, 1992; Laitman *et al.* 1979; Lieberman 1984; Trinkaus 1989). All of these, particularly when found in mid-latitude Europe, imply that cultural solutions replaced physical adaptation in dealing with cold and scarcity. Other factors may also play a role: reduced sexual dimorphism may imply a change in male-female relationships, a decreased selection for male robusticity, or simply decreased stress on females during their growth period. It is interesting also that in European specimens, Harris' lines and enamel hypoplasias decrease in early *Homo sapiens sapiens* compared to

Neandertals, but wear on deciduous teeth increases, implying either slower growth or earlier weaning and greater potential for population growth (Skinner 1989). Among the "Cro-Magnons,"[5] we find the first "old" skeletons with the possible exception of the old man from La Ferrassie (Trinkaus & Tompkins 1990; Trinkaus 1993), the only Neandertal known (or thought) to have reached the age of 50. In addition to superior technology and social complexity, old age might be another factor which increased survivorship of the young and thus the population growth potential of *H. s. sapiens* in relation to Neandertals.

The oldest *H. s. sapiens* fossils, dating to 80-130 Ka, are found in Africa at the sites of Klasies River Mouth, Border Cave, Equus Cave, Sea Harvest, and Die Kelders in South Africa, Mumba in Tanzania, Omo (Kibish Formation) in Ethiopia, and possibly Jebel Irhoud in Morocco (Bräuer 1984; Beaumont 1980; Singer & Wymer 1982; Grine & Klein 1985, 1993; Miller *et al.* 1992; Grine *et al.* 1991; Mehlman 1979, 1987; Hublin 1992; Brooks *et al.* 1993). These are followed closely by Qafzeh in the Near East (Valladas *et al.* 1988; Schwarcz *et al.* 1988; Vandermeersch 1981), although one could argue that Qafzeh was zoogeographically part of Africa at that moment (Tchernov 1988). Some authors have argued, based primarily on the limb bones from Klasies River Mouth, that these early specimens should be attributed to an "archaic" *H. sapiens* group instead (Smith 1994; Holliday *et al.* 1993). If the Border Cave dates for the jaw (BC 5) and infant burial (BC 3) are sustained, however, the evidence for gracilization in the face and cranium at an early date seems indisputable (Miller *et al.* 1992).

The behavioral archaeology of *H. s. sapiens* suggests major shifts in human adaptation in five spheres:

1. *Distribution*: Anatomically modern humans first appear in Africa and the Levant, then in East Asia and Australia (Jones 1992), and finally in Europe and Siberia by just before ca. 40 Ka (Hoffecker *et al.* 1993; Goebel *et al.* 1993). Apparently the modern human form could not or did not dominate or replace the Neandertal one until many thousands of years after the first appearance of the former. Apparently, modern humans also were the first human occupants of high latitudes (north European Plain, eastern Siberia) by 20 Ka, and possibly also of the tropical forest (Bailey *et al.* 1989; Brooks & Robertshaw 1990).

2. *Tools*: Modern humans are characterized by new technologies—blades, burins, backed pieces, bone tools, hafting, and projectiles such as spears, atlatls, harpoons, and the bow and arrow (Petersen *et al.* 1993). Raw materials may come from more than 100 miles away (White 1990) and be derived from specialized production or mining. There is more specialized use of fire to heat-treat stone, make fired clay, or process pigments for painting (Vandiver *et al.* 1989). Boats, although not recovered directly in the archaeological record, were probably what allowed dispersal to islands and island continents (Allen *et al.* 1988; Jones 1992).

3. *Diet*: Faunal remains of single species, in a limited season, with prime age animals dominating, suggest more ability to plan ahead and procure particular animals by intercept or ambush hunting (Stiner 1991, 1994; Kuhn & Stiner 1994). More dangerous animals were hunted (Klein 1978, 1989). Evidence also exists for trapping (warthogs, boars, or hares), fishing, and netting (birds), as well as for seasonal consumption of resources such as snails, seabird eggs, and seal pups (Troeng 1993). There is less carnivore overlay on the faunal remains, implying longer occupation of sites (Kuhn & Stiner 1994). The greater frequency of small animals in the African record may also imply longer occupation of sites, as these tend to be completely consumed by animal scavengers if fresh when the humans leave (Yellen 1991).

4. *Social organization*: Modern humans clearly had larger, more complex social ties, indicated by regional differentiation in stone tool styles, long distance trade, use of ornaments to indicate status, and differences in status and wealth by 20 Ka (Soffer 1985, 1992). Artistic and decorative traditions develop (Marshack 1989a, 1989b). All these may have related to the development of social networks as answers to risk and scarcity (Wiessner 1983, 1984). The elaboration of social status differences appears greatest in the coldest regions (Russia, Siberia), where risk and scarcity are greatest.

5. *Symbolic behavior*: This is exhibited in art, beads, and pendants, and in decoration of both rock walls and utilitarian objects (Bednarik 1992). Notched and engraved pieces suggest the beginnings of notation and even mathematics (Marshack 1989a, 1989b; de Heinzelin 1961; Brooks & Smith 1987). The iconic or "isochrestic" value of style may serve to link much larger social entities as well as to establish their frontiers against the incursions of others (Hodder 1985; Sackett 1980, 1982). Their presence may thus imply the existence of such large-scale groups, whose members may meet one another as rarely as once in a lifetime if at all. In addition, as in Australian aboriginal painting and carving, images may encode group experiences and serve as maps of space, time, and lifeways, outliving the memory of any one individual. Such long-term memories may be particularly valuable during the one year in a century when the reindeer or the rains do not arrive.

Do the behavioral and morphological transitions to anatomically modern *Homo* occur at the same time? Although this appeared to be the case in Europe, the evidence from the Near East and Africa suggested that these transitions did not coincide (Clark 1988, 1992). In recent scenarios modern humans were thought to have appeared but to have lived very much like Neandertals until they developed enough technological expertise and social and symbolic elaboration to take Europe from the Neandertals (Klein 1989, 1995). Now, more and more evidence from Africa suggests that the "revolution" began there about when modern humans did, or not long thereafter. By 80 Ka technologies included blades (Clark 1988; Phillipson 1993), hafted projectiles (Brooks & Yellen 1987), and bone tools (Volman 1984; Brooks *et al.* 1995; Yellen *et al.* 1995). Economies involved fishing and seasonally-specific ambush hunting of dangerous animals, *e.g.* warthogs and giant buffalo (Brooks *et al.* 1980; Brooks *et al.* 1995; Yellen *et al.* 1995; Wendorf *et al.* 1993), as well as long distance procurement of raw materials (Vermeersch 1990; Beaumont 1973, 1992; Merrick & Brown 1984; Merrick *et al.* 1994). Seasonal hunting may also occur at Qafzeh (Lieberman 1993; Lieberman & Shea 1994). In Africa, social networks are suggested by regional differences in projectile style (Clark 1988) and by movement of raw materials (Ambrose & Lorenz 1990), while incipient symbolic behavior is revealed in the few ornaments (*e.g.*, the *Conus* shell in the Border Cave burial), and decorated items such as the incised ostrich eggshells from Apollo 11, Namibia and Diepkloof, South Africa (Beaumont 1992), as well as by ground and processed pigments in a number of sites such as ≠ Gi and Kalkbank (Brooks & Yellen 1987; Mason 1988a). The oldest representative paintings, however, appear much later, in the same general time range as those of western Europe and Australia (Wendt 1972).

Symbolic activities begin to expand at ca. 50 Ka, just before the first appearance of modern humans in Europe. We now have ostrich eggshell beads from several sites in this time range in central/eastern Africa such as Enkapune ya Muto (Ambrose 1994) and Mumba (Hare *et al.* 1993), as well as possibly in southern Africa at Bushman Rock shelter, Cave of Hearths, and Border Cave (Wadley 1993). A major change in technology also takes place at or before 50 Ka in at least some regions of Africa, with the introduction of microlithic and small bipolar technology. This must have involved composite tools such as arrows or drills (Van Noten 1977; Brooks & Robertshaw 1990; Mehlman & Brooks 1992; Deacon & Geleijnse 1988). The first "Upper Paleolithic" of both North Africa and Near East, just before 40 Ka, also involves small bladelets, not large crude blades, steep scrapers, and bone points as in Europe (Bar-Yosef 1994). Travel to Australia at or before this time implies the existence of boats (Roberts *et al.* 1990). The last 20,000 years are characterized by increasing manipulation of the environment itself through burning and possible herd management, long before actual domestication and agriculture (Mellars 1976).

Continuity or Replacement?

Can archaeology tell us about how and why moderns replaced Neandertals? "Why" is easier than "how". Moderns may have had shorter birth intervals and better assurance of survivorship by "grandparents" as well as by survival of parents into their children's adulthood. Moderns also exhibited better hunting techniques, alternative food sources (fish, snails), broad social networks and the ability to interact over long distances, increasing their chances of finding enough to eat if the local resources failed. Whether these new behaviors were the consequence or the cause of biological gracilization and brain evolution is as yet unclear. The new African evidence suggests that the "great leap forward" may actually consist of a number of crabwise steps, coinciding with a mosaic and gradual achievement of anatomical modernity.

"How" the change occurred, whether by replacement or by continuous change in each region, is even harder to determine (Smith 1992; Aiello 1993). The replacement theory implies the existence of a species boundary between Neandertals and modern humans which did not permit genetic mixing. In our behaviorist model, this theory also demands a strong association between modern behavior and morphology vs. archaic behavior and morphology, at least for the last 50-60 k.y. Some aspects of modern human culture, however, were shared with Neandertals by as early as 55 Ka (Lévèque *et al.* 1993; Kuhn 1995; Stiner 1994; Kuhn & Stiner 1994; Gamble 1994), implying either parallel development or behavioral diffusion across subspecies or species boundaries. And early modern culture, when it appears in the Neandertal regions, is very different from what we find in Africa. Indeed, this early European modern culture (the Aurignacian) appears in Europe as much as 6,000 years before we find it in the Near East (Bar-Yosef 1992; Bar-Yosef & Belfer-Cohen 1993, in press; Bar-Yosef *et al.* 1996). Fossils associated with these earliest Aurignacian assemblages are few and fragmentary, so that at the very least the association in Europe between the appearance of modern humans and that of modern human behavior is undocumented. On the one hand, distal limb segments of the early moderns in Europe are long, relative to proximal limb segments, implying a tropical origin according to Allen's Rule (Trinkaus 1981). On the other hand, at Vindija (Wolpoff *et al.* 1981; Smith & Ahern 1994) fragmentary cranial remains from the earliest Aurignacian assemblage have been assigned to *H. s. neanderthalensis*! Although most archaics exhibit different behaviors from most moderns, the evidence for sharp behavioral discontinuities at the temporal species boundary is ambiguous.

While the African record implies the earlier evolution of anatomically and behaviorally modern humans from archaic antecedents on that continent, setting the stage for the replacement scenario, the European record suggests that modern Europeans evolved there behaviorally from more archaic antecedents, whether indigenous or not. This transformation could have taken place as an entirely indigenous Neandertal development, a local translation of borrowed ideas about projectiles, hunting, and risk avoidance, which dramatically reduced the selection for robusticity (Brace 1992). Given the morphological discontinuity between Neandertals and early moderns, however, a more likely process might have involved local cultural evolution by a small group of anatomically modern *Homo sapiens* who made it into Europe, presumably from the Near East, around 55 Ka, before the Upper Paleolithic "revolution" (see Gamble 1994). Arriving in eastern Europe, this small group may have borrowed some ice-age adaptations from the Neandertals, possibly interbred with them to some degree, and developed rapidly in isolation to the point where they wiped out or outcompeted the Neandertals. In any case, the possibility of a major migration from Africa to Europe at 40 Ka finds little support in the archaeological record.[6]

A cautionary tale, however, is provided by the widespread appearance of anatomically modern humans in North America around 12 Ka (*e.g.*, Frison 1993). Here the derivation from northeast Asia is very clear on the basis of dental and other morphologies (Turner 1983, 1986, 1989), but obscure archaeologically since the Clovis projectile point tradition has no Siberian counterpart (Goebel *et al.* 1993; Goebel & Aksenov 1995). There is, however, very sparse and scattered evidence, particularly in Alaska, of an older, less specialized, pre-Clovis industry (Goebel *et al.* 1991; Hoffecker *et al.* 1993) with parallels in Asia. Colonization of a new environment may be in itself as strong an impetus to cultural innovation as it is to morphological diversification or adaptive radiation and demic expansion. However, in the behavioral record conservative traits, analogous to dental cusp and enamel patterns which point to ancestral relationships, are either absent or undefined in our current methodologies.

Behavior and Morphology in Human Evolution

In summary, the record of human behavioral evolution has been much enriched by new types of study: of raw material sources, use-wear, bone breakage, seasonality of prey death, and regional site structure, as well as from an integration of behavioral morphology and archaeology. The fossils and the archaeology together suggest three major behavioral shifts: (1) the first appearance of sites and tools at *ca.* 2.5-2.1 Ma; (2) the emergence of more human patterns of growth, care of adolescents, meat-procurement, hunting, use of fire, and niche expansion to Asia at *ca.* 1.6-1.4 Ma; (3) the emergence of a fully human way of life involving technological innovation, complex social networks, think-ahead hunting and general long-range planning and procurement, symbolic expression, and worldwide niche expansion between 100 and 40 Ka.

TABLE 4. Species Boundaries in the Lineage of *Homo*

Species	Discontinuity with Predecessor	
	Morphological	Behavioral
H. habilis	++ (?)	++ (?)
H. erectus	++	+++
H. sapiens (archaic)	+(?)	??
H. sapiens sapiens	+	+++

While there are also three generally recognized species of *Homo*, only two of these behavioral shifts correspond to possible species boundaries (Table 4). A major species boundary (*H. erectus/H. sapiens*) involves little recognizable change in behavior other than habitat expansion to Europe, while the major behavioral change, in current thinking, involves a subspecies rather than a species boundary. This discrepancy suggests the need for taxonomic revision in later hominid evolution, as well as for new approaches to understanding human behavioral change.

Endnotes

1. "To these anatomical traits [of *Homo habilis*] could be added the very strong evidence that was emerging from M. D. Leakey's careful excavation of occupation floors at Olduvai, that the Olduvai gracile hominids were the makers of the tools of the Oldowan industry . . . Since no convincing associations of stone artefacts with australopithecines were known, such lithocultural activities, it seemed to Napier and myself (1964), provided powerful ethological evidence, supportive of the probable presence of *Homo*" . . . "Our importing of the cultural evidence more strongly into the scale pans in weighing up the generic status of *Homo habilis* was entirely in keeping with accepted procedure that ethological evidence may be added to morphological evidence in the assessment of the systematic status of a group." (Tobias 1991:30, 38)

2. Recently, Wood (1984, 1991, 1992a, 1992b, 1993, and this volume), and Turner and Chamberlain (1989) have proposed to recognize several new species of the genus *Homo*, based on morphological variation. Specimens grouped by Tobias (1991) within *H. habilis* are divided into *H. habilis* and *H. sp. nov.* or *H. rudolfensis*, while some specimens of what was formerly recognized as African *Homo erectus* are placed in another species, *H. ergaster*. Cultural or behavioral differences between *H. habilis* and *H. rudolfensis* have not been documented. Similarly, no clear behavioral differences exist between *H. erectus* and *H. ergaster*. For the purposes of this paper, therefore, *H. habilis* and *H. erectus* may be regarded as single taxa, which may have included separate morphotypes sharing similar behaviors. Indeed until behavioral or ecological differences between the included morphotypes are demonstrated, the argument for according these morphotypes the status of separate species is in question (see discussion of "archaic" *H. sapiens*, below).

3. Note, however, that the sites most often cited in textbooks as examples of *Homo erectus* hunting behavior, Torralba and Ambrona in Spain, are too late in the Pleistocene for *Homo erectus*, at least in Europe, and probably include the activity residues of some form of archaic *Homo sapiens*.

4. Geographically this site is better seen as an extension of the Near East, rather than as evidence for early settlement of the European landmass. The date derives from a normally-po-

larized underlying tuff dated to 1.6-1.8 Ma; the interval between deposition of the tuff and deposition of the jaw, however, is uncertain. Photographs of the excavation suggest the possible presence of soil horizons and/or unconformities between the tuff and the jaw; these suggest the need for further geological and geomorphological inquiry at this important site.

5. This name derives from a group of fossils of modern but robust type excavated in the nineteenth century by Lartet and Christy from a rock-shelter in Les Eyzies, France. The stratigraphy of this excavation was poorly controlled, the deposits were entirely removed, and the site is now occupied by a hotel. Both the recovered artifacts and the very careful excavation during 1953-1964 of the Pataud site, a few hundred yards to the west, suggest that the Cro-Magnon site may have conserved evidence from almost the entire range of the early Upper Paleolithic (*ca.* 34,000-21,000 BP). Attribution of the fossils to the earliest levels, or more specifically to one of the later Aurignacian horizons, is on the basis of the excavation description and associated artifacts which contained no Gravettian forms (Movius 1971, 1995).

6. North Africa outside the Nile Valley appears less likely as a direct staging point for the first incursions into Europe by anatomically modern humans. Much of North Africa was very sparsely settled prior to 40 Ka (Wendorf *et al.* 1993), and the distinctive Aterian projectile points and tanged scrapers of the North African Middle Stone Age have no counterparts in Europe. Neither Sicily nor southern Spain, two of the more logical points for such a crossing, have particularly early manifestations of the Upper Paleolithic. The last Mousterian sites of southern Spain are almost 15,000 years later than the earliest Aurignacian sites of eastern Europe (Straus 1994; Straus *et al.* 1993).

Literature Cited

Aiello, L. C. 1993. The fossil evidence for modern human origins in Africa: A revised view. *American Anthropol.* 95:73-96.

_____ , & P. L. Wheeler. 1995. The expensive tissue hypothesis: The brain and the digestive system in human and primate evolution. *Curr. Anthropol.* 36:199-221.

Albrecht, G. H., & J. M. A. Miller. 1993. Geographical variation in primates: A review with implications for interpreting fossils. Pages 123-161 *in* W. H. Kimbel & L. B. Martin, eds., *Species, Species Concepts, and Primate Evolution.* Plenum Press, New York.

Allen, J., C. Gosden, R. Jones, & T. P. White. 1988. Pleistocene dates for the human occupation of New Ireland, northern Melanesia. *Nature* 331:707-709.

Ambrose, S. H. 1994. Technological change: Volcanic winter, and genetic evidence for the spread of modern humans in Africa. Paper presented at the annual meeting of the Palaeoanthropology Society, April 19-20, Anaheim, California.

_____ , & K. G. Lorenz. 1990. Social and ecological models for the Middle Stone Age in southern Africa. Pages 3-33 *in* P. Mellars, ed., *The Emergence of Modern Humans.* Edinburgh University Press, Edinburgh.

Arensberg, B., L. A. Schepartz, A.-M. Tillier, B. Vandermeersch, & Y. Rak. 1990. A reappraisal of the anatomical basis for speech in Middle Palaeolithic hominids. *American Jour. Phys. Anthropol.* 83:137-146.

Arsuaga, J. L., I. Martínez, A. Gracia, J.-M. Carretero, & E. Carbonell. 1993. Three new human skulls from the Sima de los Huesos Middle Pleistocene site in Sierra de Atapuerca, Spain. *Nature* 362:534-537.

Avery, G. 1984. Sacred cows or jackal kitchens, hyaena middens and bird nests: Some implications of multi-agent contributions to archaeological accumulations. Pages 344-348 *in* M. Hall, G. Avery, D. M. Avery, M. L. Wilson, & A. J. B. Humphreys, eds., *Frontiers: Southern African Archaeology Today*. B. A. R. International Series 207, Oxford.

Bahn, P. 1996. Treasure of the Sierra Atapuerca. *Archaeology* 49(1):45-48.

Bailey, R. C., G. Head, M. Jenike, B. Owens, R. Rechtman, & E. Zechanter. 1989. Hunting and gathering in tropical rain forests: Is it possible? *American Anthropol.* 91:59-82.

Bar-Yosef, O. 1991. Les premiers peuplements humains du proche-orient. Page 235 *in* E. Bonifay & B. Vandermeersch, eds., *Les Premiers Européens*. Éditions du C. N. R. S., Paris, France.

_____. 1992. The role of western Asia in modern human origins. *Philos. Trans. R. Soc. London,* ser. B, 337:193-200.

_____. 1994. The contributions of Southwest Asia to the study of the origin of modern humans. Pages 121-132 *in* M. H. Nitecki & D. V. Nitecki, eds., *Origins of Anatomically Modern Humans*. Plenum Press, New York.

_____, M. Arnold, A. Belfer-Cohen, P. Goldberg, R. Housley, H. Laville, L. Meignen, N. Mercier, J. C. Vogel, & B. Vandermeersch. 1996. The dating of the Upper Paleolithic layers in Kebara cave, Mt. Carmel. *Jour. Archaeol. Sci.* 23:297-306.

_____, & A. Belfer-Cohen. 1993. The Levantine Aurignacian: An interim report. Pages 277-282 *in* L. Banesz, I. Cheben, L. Kaminska, & V. Pavukova, eds., *Actes du XIIe Congrès International des Sciences Préhistoriques et Protohistoriques, Bratislava, Czechoslovakia*.

_____. (in press). Another look at the Levantine Aurignacian. *Acts of the XIII Congress of Pre- and Protohistoric Sciences*. Colloquium X, The Origin of Modern Man, September 8-14, 1996, Forli, Italy.

Beaumont, P. B. 1973. The ancient pigment mines of southern Africa. *South African Jour. Sci.* 69:140-146.

_____. 1980. On the age of Border Cave hominids 1-5. *Palaeontol. Afr.* 23:21-33.

_____. 1992. The time depth of aesthetic and symbolic behavior in southern Africa. *South African Assoc. Archaeol. Biennial Conf., Abstracts*, p. 39.

Bednarik, R. G. 1992. Oldest dated rock art: a revision. *The Artefact* 15:39.

Begun, D., & A. Walker. 1993. The endocast. Pages 326-358 *in* A. Walker & R. Leakey, eds., *The Nariokotome* Homo erectus *Skeleton*. Harvard University Press, Cambridge, Massachusetts.

Bellomo, R. V. 1993. A methodological approach for identifying archaeological evidence of fire resulting from human activities. *Jour. Archaeol. Sci.* 20:525-555.

_____. 1994. Early Pleistocene fire technology in northern Kenya. Pages 16-28 *in* S. T. Childs, ed., *Society, Technology and Culture In Africa*. MASCA Research Papers in Science and Archaeology 11 (Supplement), University of Pennsylvania Press, Philadelphia, Pennsylvania.

Binford, L. R. 1979. Organization and formation processes: Looking at curated technologies. *Jour. Anthropol. Res.* 33:493-502.

_____. 1980. Willow smoke and dogs' tails: Human settlement systems and archaeological site formation. *American Antiquity* 45:4-20.

_____. 1982. The archaeology of place. *Jour. Anthropol. Archaeol.* 1:5-31.

_____. 1983. Long-term land-use patterns:Some implications for archaeology. Pages 379-386 *in* L. R. Binford, ed., *Working at Archaeology*. Academic Press, New York.

_____. 1985. Human ancestors: Changing views of their behavior. *Jour. Anthropol. Archaeol.* 4:292-327.

_____. 1987. Searching for camps and missing the evidence: Another look at the Lower Palaeolithic. Pages 17-31 *in* O. Soffer, ed., *The Pleistocene Old World: Regional Perspectives*. Plenum Press, New York.

_____. 1989. Isolating the transition to cultural adaptations: An organizational approach. Pages 18-41 in E. Trinkaus, ed., *The Emergence of Modern Humans: Biocultural Adaptations in the Later Pleistocene*. Cambridge University Press, Cambridge, UK.

_____, & S. R. Binford. 1966. A preliminary analysis of functional variability in the Mousterian of Levallois facies. *American Anthropol.* 68:238-295.

_____, & C. K. Ho. 1985. Taphonomy at a distance: Zhoukoudian, "The Cave Home of Beijing Man"? *Curr. Anthropol.* 26:413-442.

_____, & N. M. Stone. 1986. Zhoukoudian: A closer look. *Curr. Anthropol.* 27:453-576.

Blumenschine, R., J. A. Cavallo, & S. D. Capaldo. 1994. Competition for carcasses and early hominid behavioral ecology: A case study and a conceptual framework. *Jour. Hum. Evol.* 27:197-213.

Bordes, F. 1972. *A Tale of Two Caves*. Harper & Row, New York. 169 pp.

Brace, C. L. 1964. The fate of the classic Neanderthals: A consideration of hominid catastrophism. *Curr. Anthropol.* 5:3-43.

_____. 1992. *Modern Human Origins: Narrow Focus or Broad Spectrum*. David Skomp Distinguished Lectures in Anthropology, Indiana University, April 16.

_____. 1995. Biocultural interaction and the mechanism of mosaic evolution in the emergence of "modern" morphology. *American Anthropol.* 97:711-721.

_____, A. S. Ryan, & B. H. Smith. 1981. Comment on "Tooth wear in La Ferrassie Man" by Pierre-François Puech. *Curr. Anthropol.* 22:426-430.

Brain, C. K. 1968. Who killed the Swartkrans ape-men? *South African Mus. Assoc. Bull.* 9:127-139.

_____. 1981. *The Hunters or the Hunted? An Introduction to African Cave Taphonomy*. University of Chicago Press, Chicago, Illinois. 365 pp.

_____, & A. Sillen. 1988. Evidence from the Swartkrans cave for the earliest use of fire. *Nature* 336:464-466.

Bräuer, G. 1984. The "Afro-European *sapiens* hypothesis" and hominid evolution in East Asia during the late Middle and Upper Pleistocene. *Cour. Forschungsinst. Senckenb.* 69:145-165.

Bromage, T. G. 1987. The biological and chronological maturation of early hominids. *Jour. Hum. Evol.* 16:257-272.

Brooks, A. S., A. L. Crowell, & J. E. Yellen. 1980. ≠ Gi: A stone age archaeological site in the northern Kalahari desert, Botswana. Pages 304-309 in R. E. F. Leakey & B. A. Ogot, eds., *Proceedings of the Eighth Panafrican Congress of Prehistory and Quaternary Studies*. The International Louis Leakey Memorial Institute for African Prehistory, Nairobi.

Brooks, A. S., & C. C. Smith. 1987. Ishango revisited: New age determination and cultural interpretations. *Afr. Archaeol. Rev.* 5:65-78.

Brooks A. S., & J. E. Yellen. 1987. The preservation of activity areas in the archaeological record: Ethnoarchaeological and archaeological work in northwest Ngamiland, Botswana. Pages 63-106 in S. Kent, ed., *Method and Theory of Activity Area Research: An Ethnoarchaeological Approach*. Columbia University Press, New York.

Brooks, A. S., & P. Robertshaw. 1990. The glacial maximum in tropical Africa 22,000-12,000 BP. Pages 121-169 in C. S. Gamble & O. Soffer, eds., *The World at 18,000 BP, Volume 2: Low Latitudes*. Unwin Hyman, London, UK.

Brooks, A. S., & B. A. Wood. 1990. The Chinese side of the story. *Nature* 344:288-289.

Brooks, A. S., P. E. Hare, & J. E. Kokis. 1993. Age of early anatomically modern human fossils from the cave of Klasies River Mouth, South Africa. *Carnegie Inst. Washington Year Book* 92:95-96.

Brooks, A. S., J. S. Cramer, A. Franklin, J. de Heinzelin, D. M. Helgren, W. Hornyak, J.

Keating, R. G. Klein, W. J. Rink, H. Schwarcz, J. N. L. Smith, K. Stewart, N. Todd, J. Verniers, & J. Yellen. 1995. Dating and context of three Middle Stone Age sites with bone artifacts in the Upper Semliki Valley, Zaire. *Science* 268:548-553.

Brooks, A. S., & G. Laden. (in press). Environmental determinants of site formation: A comparison of ethnoarchaeological work in the Kalahari desert and Ituri forest, with implications for the "emergence" of human culture. *Proceedings of the Xth Pan-African Congress of Prehistory and Related Studies, Harare, Zimbabwe, June 25-30, 1995.* Government Printers, Harare, Zimbabwe.

Bunn, H., & E. Kroll. 1986. Systematic butchery by Plio-Pleistocene hominids at Olduvai Gorge, Tanzania. *Curr. Anthropol.* 27:431-452.

Carbonell, E., J. M. Bermúdez de Castro, J. L. Arsuaga, J. C. Diez, A. Rosas, G. Cuenca-Bescós, R. Sala, M. Mosquera, & X. P. Rodríguez. 1995. Lower Pleistocene hominids and artifacts from Atapuerca-TD6 (Spain). *Science* 269:826-830.

Cerling, T. 1992. Development of grasslands and savannas in East Africa during the Neogene. *Palaeogeogr. Palaeoclimatol. Palaeoecol.* 63:335-356.

Chavaillon, J. 1976. Evidence for the technical practices of early Pleistocene hominids: Shungura Formation, lower Omo Valley, Ethiopia. Pages 565-573 *in* Y. Coppens, F. C. Howell, G. L. Isaac & R. E. F. Leakey, eds., *Earliest Man and Environments in the Lake Rudolf Basin: Stratigraphy, Paleoecology, and Evolution.* University of Chicago Press, Chicago, Illinois.

Clark, J. D. 1970. *The Prehistory of Africa.* Thames & Hudson, London, UK. 302 pp.

_____. 1988. The Middle Stone Age of East Africa and the beginnings of regional identity. *Jour. World Prehist.* 2:235-303.

_____. 1992. African and Asian perspectives on the origins of modern humans. *Philos. Trans. R. Soc. London,* ser. B, 337:201-215.

_____, J. de Heinzelin, K. Schick, W. K. Hart, T. D. White, G. WoldeGabriel, R. C. Walter, G. Suwa, B. Asfaw, & E. Vrba. 1994. African *Homo erectus*: Old radiometric ages and young Oldowan assemblages in the Middle Awash Valley, Ethiopia. *Science* 264:1907-1910.

Clarke, R. J. 1990. The Ndutu cranium and the origin of *Homo sapiens. Jour. Hum. Evol.* 19:699-736.

Darwin, C. 1859. *On the Origin of Species by Means of Natural Selection.* John Murray, London, UK. 513 pp.

Deacon, H. J., & V. B. Geleijnse. 1988. The stratigraphy and sedimentology of the main site sequence: Klasies River, South Africa. *South African Archaeol. Bull.* 43:5-14.

Dean, M. C. 1990. More on cool heads: Reply to Wheeler. *Jour. Hum. Evol.* 19:323-325.

Defleur, A. 1993. *Les Sépultures Moustériennes.* Éditions du CNRS, Paris, France. 325 pp.

_____, O. Dutour, H. Valladas, & B. Vandermeersch. 1993. Cannibals among the Neanderthals? *Nature* 362:214.

Delson, E. 1991. Combien d'espèces de genre *Homo* existe-t-il en Europe. Page 283 *in* E. Bonifay & B. Vandermeersch, eds., *Les Premiers Européens.* Éditions du CNRS, Paris, France.

_____, N. Eldredge, & I. Tattersall. 1977. Reconstruction of hominid phylogeny: A testable framework based on cladistic analysis. *Jour. Hum. Evol.* 6:263-278.

D'Errico, F., & P. Villa. 1996. Holes and grooves: The contribution of anatomy and taphonomy to the problem of art origins. Paper [oral] presented at the annual meeting of the Palaeoanthropology Society, April 9-10, New Orleans, Louisiana.

Diamond, J. 1992. *The Third Chimpanzee: The Evolution and Future of the Human Animal.* Harper Collins, New York. 407 pp.

Dobzhansky, T. 1937. *Genetics and the Origin of Species*. Columbia University Press, New York. 364 pp.

Dzaparidze, V., G. Bosinski, T. Bugianisvili, L. Gabunia, A. Justus, N. Klopotovskaja, E. Kvavadze, D. Lordkipanidze, G. Majsuradze, N. Mgeladze, M. Nioradze, E. Pavlenisvili, H.-U. Schmincke, D. Sologasvili, D. Tusabramisvili, M. Tvalcrelidze, & A. Vekua. 1992. Der altpaläolithische Fundplatz Dmanisi in Georgien (Kaukasus). *Jahrb. Röm.-German. Zentralmus. Mainz* 36:67-116.

Eldredge, N. 1986. Information, economics and evolution. *Annu. Rev. Ecol. Syst.* 17:351-369.

_____. 1993. What, if anything, is a species? Pages 3-20 *in* W. H. Kimbel & L. B. Martin, eds., *Species, Species Concepts, and Primate Evolution*. Plenum Press, New York.

Falk, D. 1986. Evolution of cranial blood drainage in hominids: Enlarged occipital/marginal sinuses and emissary foramina. *American Jour. Phys. Anthropol.* 70:311-324.

_____. 1987a. Brain lateralization in primates and its evolution in hominids. *Yearb. Phys. Anthropol.* 30:107-125.

_____. 1987b. Hominid palaeoneurology. *Annu. Rev. Anthropol.* 16:13-30.

_____. 1990. Brain evolution in *Homo*: The radiator theory. *Behav. Brain Sci.* 13:333-381.

_____. 1992. *Braindance*. Henry Holt, New York. 260 pp.

Fogel, M., N. Tuross, & D. Owsley. 1989. Nitrogen isotope traces of human lactation in modern and archaeological populations. *Annu. Rep. Director Geophys. Lab. Carnegie Inst. Washington* 1988-89:111-117.

Foley, R. 1987. *Another Unique Species: Patterns in Human Evolutionary Ecology*. Longman, Harlow, UK. 313 pp.

Frayer, D., & M. Wolpoff. 1985. Human sexual dimorphism. *Annu. Rev. Anthropol.* 14:429-473.

Frison, G. C. 1993. *Prehistoric Hunters of the High Plains*. Academic Press, New York. 457 pp.

Gabunia, L., & A. Vekua. 1995. A Plio-Pleistocene hominid from Dmanisi, East Georgia, Caucasus. *Nature* 373:509-512.

Gamble, C. S. 1994. *Timewalkers: The Prehistory of Global Colonization*. Harvard University Press, Cambridge, Massachusetts. 309 pp.

Gargett, R. 1989. The evidence for Neandertal burial. *Curr. Anthropol.* 30:157-190.

Goebel, T., N. Bigelow, & R. Powers. 1991. The Nenana complex of Alaska and Clovis origins. Pages 49-79 *in* R. Bonnischen & K. Turnmire, eds., *Clovis Origins and Adaptations*. Center for the Study of the First Americans, Corvallis, Oregon.

Goebel, T., A. P. Derevianko, & V. T. Petrin. 1993. Dating the Middle-to-Upper Palaeolithic transition at Kara-Bom. *Curr. Anthropol.* 34:452-458.

Goebel, T., & M. Aksenov. 1995. Accelerator radiocarbon dating of the initial Upper Paleolithic in southeast Siberia. *Antiquity* 69:349-357.

Gordon, K. 1993. Reconstructing hominid diet in the Plio-Pleistocene. *Riv. Antropol. Roma* 71:71-89.

Grine, F. E., & R. G. Klein. 1985. Pleistocene and Holocene human remains from Equus Cave, South Africa. *Anthropology* 8:55-98.

_____. 1993. Late Pleistocene human remains from the Sea Harvest site, Saldanha Bay, South Africa. *South African Jour. Sci.* 89:145-152.

Grine, F. E., R. G. Klein, & T. P. Volman. 1991. Dating, archaeology and human fossils from the Middle Stone Age levels of Die Kelders, South Africa. *Jour. Hum. Evol.* 21:363-395.

Hare, P. E., G. A. Goodfriend, A. S. Brooks, J. E. Kokis, & D. W. von Endt. 1993. Chemical clocks and thermometers: Diagenetic reactions of amino acids in fossils. *Carnegie Inst. Washington Year Book* 92:80-85.

Hayden, B. 1993. The cultural capacities of Neandertals: A review and re-evaluation. *Jour. Hum. Evol.* 24:113-146.

Haynes, G. 1981. *Bone Modifications and Skeletal Disturbances by Natural Agencies: Studies in North America.* Ph.D. Dissertation, Catholic University of America, Washington, D. C.

Heinzelin, J. de. 1961. Ishango. *Sci. American* 206:105-116.

Hill, A., S. Ward, A. Deino, G. Curtis, & R. Drake. 1992. Earliest *Homo. Nature* 355:719-722.

Hodder, I. 1985. *Symbols in Action: Ethnoarchaeological Studies of Material Culture.* Cambridge University Press, Cambridge, UK. 244 pp.

Hoffecker, J. F., W. R. Powers, & T. Goebel. 1993. The colonization of Beringia and the peopling of the New World. *Science* 259:46-53.

Holliday, T. W., S. E. Churchill, & E. Trinkaus. 1993. Modern human origins in Africa: The postcranial evidence. Paper [oral] presented at the annual meeting of the American Association for the Advancement of Science.

Holliday, T. W., & E. Trinkaus. 1991. Limb-trunk proportions in Neandertals and early anatomically modern humans. *American Jour. Phys. Anthropol.* Suppl. 12:93-94.

Huang, W., R. Ciochon, Y. Gu, R. Larick, Q. Fang, H. Schwarcz, C. Yonge, J. de Vos, & W. Rink. 1995. Early *Homo* and associated artifacts from Asia. *Nature* 378:275-278.

Hublin, J.-J. 1992. Recent human evolution in northwest Africa. *Philos. Trans. R. Soc. London* B 337:193-200.

James, S. R. 1989. Hominid use of fire in the Lower and Middle Pleistocene. *Curr. Anthropol.* 30:1-26.

Jones, R. 1992. The human colonization of the Australian continent. Pages 289-301 *in* G. Bräuer & F. H. Smith, eds., *Continuity or Replacement: Controversies in* Homo sapiens *Evolution.* A. A. Balkema, Rotterdam, The Netherlands.

Kalb, J. E., M. Jaeger, C. P. Jolly, & B. Kana. 1982. Preliminary geology, paleontology and palaeoecology of a Sangoan site at Anderlee, Middle Awash Valley, Ethiopia. *Jour. Archaeol. Sci.* 9:349-363.

Keeley, L. H. 1980. *Experimental Determination Of Stone Tool Uses: A Microwear Analysis.* University of Chicago Press, Chicago, Illinois. 212 pp.

_____ , & N. Toth. 1981. Microwear polishes on early stone tools from Koobi Fora, Kenya. *Nature* 293:464-465.

Kennedy, G. E. 1992. The evolution of *Homo sapiens* as indicated by features of the post-cranium. Pages 209-218 *in* G. Bräuer & F. H. Smith, eds., *Continuity or Replacement: Controversies in* Homo sapiens *Evolution.* A. A. Balkema, Rotterdam, The Netherlands.

Kibunjia, M., H. Roche, F. Brown, & R. Leakey. 1992. Pliocene and Pleistocene archaeological sites west of Lake Turkana, Kenya. *Jour. Hum. Evol.* 23:431-438.

Kimbel, W. H., & L. B. Martin, eds. 1993. *Species, Species Concepts, and Primate Evolution.* Plenum Press. New York. 560 pp.

Kimbel, W. H., & Y. Rak. 1993. The importance of species taxa in paleoanthropology and an argument for the phylogenetic concept of the species category. Pages 461-484 *in* W. H. Kimbel & L. B. Martin, eds., *Species, Species Concepts, and Primate Evolution.* Plenum Press, New York.

Kimbel, W. H., D. C. Johanson, & Y. Rak. 1994. First skull and other new discoveries of *Australopithecus afarensis* at Hadar, Ethiopia. *Nature* 368:449-451.

Klein, R. G. 1978. Stone age predation on large African bovids. *Jour. Archaeol. Sci.* 5:95-217.

_____ . 1982. Age (mortality) profiles as a means of distinguishing hunted species from scavenged ones in the archaeological record. *Paleobiology* 8:151-158.

_____ . 1989. *The Human Career.* University of Chicago Press, Chicago, Illinois. 524 pp.

_____ . 1995. Anatomy, behavior and modern human origins. *Jour. World Prehist.* 9:167-198.

_____ , & K. Cruz-Uribe. 1991. The bovids from Elandsfontein, South Africa, and their implications for the age, paleoenvironment and origins of the site. *Afr. Archaeol. Rev.* 9:21-79.

Knecht, H. 1993. Early Upper Palaeolithic approaches to bone and antler projectile technology. Pages 33-47 *in* G. L. Petersen, H. M. Bricker, & P. Mellars, eds., *Hunting and Animal Exploitation in the Later Palaeolithic and Mesolithic of Eurasia*. American Anthropological Association, Washington DC.

Kuhn, S. L. 1991. "Unpacking" reduction: Lithic raw material economy in the Mousterian of west-central Italy. *Jour. Anthropol. Archaeol.* 10:76-106.

_____ . 1992. On planning and curated technologies in the Middle Palaeolithic. *Jour. Anthropol. Res.* 48:185-214.

_____ . 1995. *Mousterian Lithic Technology: An Ecological Perspective*. Princeton University Press, Princeton, New Jersey. 209 pp.

_____ , & M. C. Stiner. 1994. Behavioral ecology and human origins research. Paper [oral] presented at the annual meeting of the Society for American Archaeology, April 20-24, Anaheim, California.

Laden, G., & A. S. Brooks. 1994. The effects of the landscape on the archaeological record of foragers: Contrasting the Kalahari and the Ituri. Paper [oral] presented at the 12th Biennial Meeting of the Society of Africanist Archaeologists, April 28-May 1.

Laitman, J. T., R. C. Heimbuch, & E. S . Crelin. 1979. The basicranium of fossil hominids as an indicator of their upper respiratory systems. *American Jour. Phys. Anthropol.* 51:15-34.

Le Mort, F. 1989. Traces de décharnement sur les ossements néandertaliens de Combe-Grenal (Dordogne). *Bull. Soc. Préhist. Fr.* 86:79-87.

Leakey, L. S. B., P. V. Tobias, & J. Napier. 1964. A new species of the genus *Homo* from Olduvai Gorge. *Nature* 202:7-9.

Leakey, M. D. 1971. *Olduvai Gorge, Volume 3: Excavations in Beds I and II, 1960-1963*. Cambridge University Press, Cambridge, UK. 306 pp.

_____ . 1995. *Olduvai Gorge, Volume 5: Excavations in Beds III, IV and the Masek Beds, 1968-1971*. Cambridge University Press, Cambridge, UK. 341 pp.

Lee-Thorp, J. A. & F. Thackery. 1996. Isotopic ratios and diets of *Homo* and *Australopithecus robustus* compared at Swartkrans. Paper [oral] presented at the annual meeting of the Palaeoanthropology Society, April 9-10, New Orleans, Louisiana.

Lévèque, F., A. M. Backer, & M. Guilbaud. 1993. *Context of a Late Neandertal: Implications of Multidisciplinary Research for the Transition to Upper Paleolithic Adaptations at Saint-Césaire (Charente-Maritime), France*. Prehistory Press, Madison, Wisconsin. 131 pp.

Lévèque, F., & B. Vandermeersch. 1980. Découverte des restes humains dans un niveau castelperronian à Saint-Césaire (Charente-Maritime). *C. R. Hebd. Séances Acad. Sci.,* Ser. D, 291:187-189.

Lieberman, D. E. 1993. The rise and fall of seasonal mobility: The case of the southern Levant. *Curr. Anthropol.* 34:599-631.

_____ , & J. J. Shea. 1994. Behavioral differences between archaic and modern humans in the Levantine Mousterian. *American Anthropol.* 96:300-332.

Lieberman, P. 1984. *The Biology and Evolution of Language*. Harvard University Press, Cambridge, Massachusetts. 379 pp.

Lovejoy, C. O. 1981. The origins of man. *Science* 211:341-350.

Loy, T. 1983. Prehistoric blood residues: Detection on tool surfaces and identification of species of origin. *Science* 220:1269-1270.

Lumley, H. de 1969. A palaeolithic camp at Nice. *Sci. American* 220:42-50.

_____ . 1988. La grotte de Vallonnet. *Anthropologie* (Paris) 92:387-397.

Marshack, A. 1989a. The Neanderthals and the human capacity for symbolic thought: Cognitive and problem-solving aspects of Mousterian symbol. Pages 57-91 *in* M. Otte, ed., *L'Homme de Néanderthal 5: La Pensée*. Études et Recherches Archéologiques de l'Université de Liège, Liège, Belgium.

_____ . 1989b. Evolution of the human capacity: The symbolic evidence. *Yearb. Phys. Anthropol.* 32:1-34.

Martin, L. 1985. Significance of enamel thickness in hominoid evolution. *Nature* 314:260-263.

Mason, R. J. 1988a. A Middle Stone Age faunal site at Kalkbank, northern Transvaal: Archaeology and interpretation. Pages 201-203 *in* R. van Zinderen Bakker, ed., *Palaeoecology of Africa and the Surrounding Islands, Volume 19*. A. A. Balkema, Rotterdam.

_____ . 1988b. Cave of Hearths, Makapansgat, Transvaal. *Occas. Pap. Archaeol. Res. Unit Univ. Witwatersrand* 21. 713 pp.

Mayr, E. 1950. Taxonomic categories in fossil hominids. *Cold Spring Harbor Symp. Quant. Biol.* 15:109-118.

_____ . 1963. *Animal Species and Evolution*. Harvard University Press, Cambridge, Massachusetts. 797 pp.

McBrearty, S. 1990. The origin of modern humans. *Man* 25:129-143.

McHenry, H. M. 1982. The pattern of human evolution: Studies on bipedalism, mastication and encephalization. *Annu. Rev. Anthropol.* 11:151-173.

_____ . 1991. Sexual dimorphism in *Australopithecus afarensis*. *Jour. Hum. Evol.* 20:21-32.

Mehlman, M. 1979. Mumba-Hohle revisited: The relevance of a forgotten excavation to some current issues in East African prehistory. *World Archaeol.* 11:80-94.

_____ . 1987. Provenience, age and associations of archaic *Homo sapiens* crania from Lake Eyasi, Tanzania. *Jour. Archaeol. Sci.* 14:133-162.

_____ , & A. S. Brooks. 1992. New light from the heart of darkness. Paper [oral] presented at the biennial meeting of the Society of Africanist Archaeologists, March, Los Angeles, California.

Mellars, P. 1976. Fire ecology, animal populations and man: A study of some ecological relationships in prehistory. *Proc. Prehist. Soc.* 42:15-45.

_____ . 1989. Technological changes at the Middle-Upper Palaeolithic transition: Economic, social and cognitive perspectives. Pages 338-365 *in* P. Mellars & C. B. Stringer, eds., *The Human Revolution: Behavioural and Biological Perspectives on the Origins of Modern Humans*. Edinburgh University Press, Edinburgh, UK.

_____ . ed. 1990. *The Emergence of Modern Humans: An Archaeological Perspective*. Edinburgh University Press, Edinburgh, UK. 555 pp.

Merrick, H. V., & F. H. Brown. 1984. Obsidian sources and patterns of source utilization in Kenya and northern Tanzania: Some initial findings. *Afr. Archaeol. Rev.* 2:129-152.

Merrick, H. V., F. H. Brown, & W. Nash. 1994. Use and movement of obsidian in the Early and Middle Stone Ages of Kenya and northern Tanzania. Pages 29-44 *in* S. T. Childs, ed., *Society, Technology and Culture in Africa*. MASCA Research Papers in Science and Archaeology 11 (Supplement), University of Pennsylvania Press, Philadelphia, Pennsylvania.

Merrick, H. V., & J. P. S. Merrick. 1976. Archaeological occurrences of earlier Pleistocene age from the Shungura Formation. Pages 574-584 *in* Y. Coppens, F. C. Howell, G. Isaac, & R. E. F. Leakey, eds., *Earliest Man and Environments in the Lake Rudolf Basin: Stratigraphy, Paleoecology, and Evolution*. University of Chicago Press, Chicago, Illinois.

Miller, G. H., P. B. Beaumont, A. T. Jull, & B. Johnson. 1992. Pleistocene geochronology and palaeothermometry from protein diagenesis in ostrich eggshells: Implications for the origin of modern humans. *Philos. Trans. R. Soc. London*, ser. B, 337:149-157.

Movius, H. L. 1950. A wooden spear of third interglacial age from Lower Saxony. *Southwest. Jour. Anthropol.* 6:139-143.

_____. 1971. The Abri de Cro-Magnon, Les Eyzies (Dordogne) and the probable age of the contained burials on the basis of the evidence of the nearby Abri Pataud. *Anu. Estudios Atlanticos* 15:323-344.

_____. 1995. Inventaire analytique des sites aurignaciens et périgordiens de Dordogne: Les Eyzies: Abri de Cro-Magnon. Pages 249-254 *in* H. M. Bricker, ed., *Le Paléolithique Supérieur de L'Abri Pataud (Dordogne): Les Fouilles de H. L. Movius, Jr.* Documents d'Archéologie Française 50, Éditions de la Maison des Sciences de l'Homme, Paris, France.

Mturi, A. 1976. New hominid from Lake Ndutu, Tanzania. *Nature* 262:484-485.

Napier, J., & P. V. Tobias. 1964. The case for *Homo habilis. The Times*, London, 5 June.

O'Brien, E. 1981. The projectile capabilities of the Acheulean hand-axe from Olorgesailie. *Curr. Anthropol.* 22:76-79.

Odell, G. 1977. *The Application of Microwear Analysis to the Lithic Component of an Entire Prehistoric Settlement: Methods, Problems and Functional Reconstructions.* Ph.D. Dissertation, Harvard University, Cambridge, Massachusetts. 716 pp.

Ogilvie, M. D., B. K. Curran, & E. Trinkaus. 1989. Incidence and patterning of dental enamel hypoplasia among the Neandertals. *American Jour. Phys. Anthropol.* 79:25-41.

Peretto, C. 1991. Les gisements d'Isernia la Pineta (Molise, Italie). Pages 161-168 *in* E. Bonifay & B. Vandermeersch, eds., *Les Premiers Européens.* Éditions du CNRS, Paris, France.

Petersen, G. L., H. M. Bricker, & P. Mellars, eds. 1993. *Hunting and Animal Exploitation in the Later Palaeolithic and Mesolithic of Eurasia.* American Anthropological Association, Washington, D. C. 248 pp.

Phillipson, D. L. 1993. *African Archaeology.* Cambridge University Press, Cambridge, UK. 268 pp.

Potts, R. 1988. *Early Hominid Activities at Olduvai.* Aldine de Gruyter, New York. 396 pp.

_____. 1994. Variables vs. models of early Pleistocene hominid land-use. *Jour. Hum. Evol.* 27:7-24.

_____, P. Shipman, & E. Ingall. 1988. Taphonomy, palaeoecology and hominids of Lainyamok, Kenya. *Jour. Hum. Evol.* 17:596-674.

Price, T. D., M. J. Schoeninger, & G. J. Armelagos. 1985. Bone chemistry and past behavior: An overview. *Jour. Hum. Evol.* 14:419-447.

Rak, Y. 1990. On the difference between two pelvises of Mousterian context from the Qafzeh and Kebara Caves, Israel. *American Jour. Phys. Anthropol.* 81:323-332.

_____. 1991. Lucy's pelvic anatomy: its role in bipedal gait. *Jour. Hum. Evol.* 20:283-290.

_____. 1993. Morphological variation in *Homo neanderthalensis* and *Homo sapiens* in the Levant: A biogeographical model. Pages 523-536 *in* W. H. Kimbel & L. B. Martin, eds., *Species, Species Concepts, and Primate Evolution.* Plenum Press, New York.

_____, & B. Arensberg. 1987. Kebara 2 Neanderthal pelvis: First look at a complete inlet. *American Jour. Phys. Anthropol.* 73:227-231.

Ricklan, D. E. 1987. Functional anatomy of the hand of *Australopithecus africanus. Jour. Hum. Evol.* 16:643-664.

Rightmire, G. P. 1990. *The Evolution of* Homo erectus*: Comparative Anatomical Studies of an Extinct Human Species.* Cambridge University Press, Cambridge, UK. 260 pp.

Roberts, R. E., R. Jones, & M. A. Smith. 1990. Thermoluminescence dating of a 50,000 year old human occupation site in northern Australia. *Nature* 345:153-156.

Rodman, P. S., & H. M. McHenry. 1980. Bioenergetics and the origin of hominid bipedalism. *American Jour. Phys. Anthropol.* 52:103-106.

Roebroeks, W. 1994. Updating the earliest occupation of Europe. *Curr. Anthropol.* 38:301-305.

Rogers, M. J., C. S. Feibel, & J. W. K. Harris. 1994. Changing patterns of land use by Plio-Pleistocene hominids in the Lake Turkana Basin. *Jour. Hum. Evol.* 27:139-158.

Ruff, C. B. 1987. Sexual dimorphism in human lower limb bone structure: Relationship to subsistence strategy and sexual division of labor. *Jour. Hum. Evol.* 16:391-416.

_____. 1991. Climate and body shape in human evolution. *Jour. Hum. Evol.* 21:81-105.

_____, E. Trinkaus, & A. Walker. 1993. Postcranial robusticity in *Homo*: Temporal trends and mechanical interpretation. *American Jour. Phys. Anthropol.* 91:21-53.

_____, & A. Walker. 1993. Body size and body shape. Pages 234-265 *in* A. Walker & R. Leakey, eds., *The Nariokotome* Homo erectus *Skeleton*. Harvard University Press, Cambridge, Massachusetts.

_____, A. Walker, & E. Trinkaus. 1994. Postcranial robusticity in *Homo:* Ontogeny. *American Jour. Phys. Anthropol.* 93:35-54.

Sackett, J. R. 1977. The meaning of style in archaeology: A general model. *American Antiquity* 42:369-380.

_____. 1980. Style and ethnicity in archaeology: The case for isochrestism. Pages 32-43 *in* M. Conkey & C. Hastorf, eds., *The Uses of Style in Archaeology*. Cambridge University Press, Cambridge, UK.

_____. 1982. Approaches to style in lithic archaeology. *Jour. Anthropol. Archaeol.* 1:59-112.

Sahnouni, M. 1987. *L'Industrie sur Galets du Gisement Villafranchien Supérieur de Ain Hanech*. Office des Publications Universitaires, Algiers, Algeria. 196 pp.

Schick, K. 1994. The Movius line reconsidered: Perspectives on the early Paleolithic of eastern Asia. Pages 569-596 *in* R. S. Corruccini & R. L. Ciochon, eds., *Integrative Paths to the Past: Paleoanthropological Advances in Honor of F. Clark Howell*. Prentice-Hall, Englewood Cliffs, New Jersey.

_____, & Z. Dong. 1993. Early Paleolithic of China and eastern Asia. *Evol. Anthropol.* 2:22-35.

_____, & N. Toth. 1993. *Making Silent Stones Speak: Human Evolution and The Dawn of Technology*. Simon & Schuster, New York. 351 pp.

_____, & N. Toth. 1994. Early stone age technology in Africa: A review and case study into the nature and function of spheroids and subspheroids. Pages 429-449 *in* R. S. Corruccini & R. L. Ciochon, eds., *Integrative Paths to the Past: Palaeoanthropological Advances in Honor of F. Clark Howell*. Prentice-Hall, Englewood Cliffs, New Jersey.

Schmandt-Besserat, D. 1980. Ocher in prehistory: 300,000 years of the use of iron ore as pigment. Pages.127-150 *in* T. A. Wertime & D. Muhly, eds., *The Coming of the Age of Iron*. Yale University Press, New Haven, Connecticut.

Schoeninger, M. J. 1979. Diet and status at Chalcatzingo: Some empirical and technical aspects of strontium analysis. *American Jour. Phys. Anthropol.* 51:295-310.

_____, M. J. DeNiro, & J. Tauber. 1983. $^{15}N/^{14}N$ ratios of bone collagen reflect marine and terrestrial components of prehistoric human diet. *Science* 220:1381-1383.

Schrenk, F., T. G. Bromage, C. G. Betzler, U. Ring, & Y. M. Juwayeyi. 1993. Oldest *Homo* and Pliocene biogeography of the Malawi rift. *Nature* 365:833-836.

Schwarcz, H. P., R. Grün, B. Vandermeersch, O. Bar-Yosef, H. Valladas, & E. Tchernov. 1988. ESR dates for the human burial site of Qafzeh in Israel. *Jour. Hum. Evol.* 17:733-737.

Schwartz, J. H., I. Tattersall, & N. Eldredge. 1978. Phylogeny and classification of the primates revisited. *Yearb. Phys. Anthropol.* 21:95-133.

Sept, J. 1992. Archaeological evidence and ecological perspectives for reconstructing early hominid subsistence behavior. *Adv. Archaeol. Method Theory* 4:1-56.

Shea J. J. 1988. Spear points from the Middle Paleolithic of the Levant. *Jour. Field Archaeol.* 15:441-450.

Shipman, P. 1986. Scavenging or hunting in early hominids: Theoretical framework and tests. *American Anthropol.* 88:27-43.

_____, W. Bosler, & K. Davis. 1981. Butchering of giant geladas at an Acheulean site. *Curr. Anthropol.* 22:257-268.

_____, & A. Walker. 1989. The costs of becoming a predator. *Jour. Hum. Evol.* 18:373-392.

Sillen, A. 1981. Strontium and diet at Hayonim Cave. *American Jour. Phys. Anthropol.* 56:131-137.

_____. 1992. Strontium/calcium ratios (Sr/Ca) of *Australopithecus robustus* and associated fauna from Swartkrans. *Jour. Hum. Evol.* 23:493-516.

_____. 1993. Was *Australopithecus robustus* an omnivore? *South African Jour. Sci.* 89:71-72.

Simpson, S. W., C. O. Lovejoy, & R. S. Meindl. 1990. Hominid dental maturation. *Jour. Hum. Evol.* 19:285-287.

Singer, R., & J. Wymer. 1968. Archaeological investigations at the Saldanha skull site in South Africa. *South African Archaeol. Bull.* 35:63-74.

_____. 1982. *The Middle Stone Age at Klasies River Mouth in South Africa*. University of Chicago Press, Chicago, Illinois. 234 pp.

Skinner, M. 1981. Dental attrition in immature hominids of the late Pleistocene: Implications for adult longevity. *American Jour. Phys. Anthropol.* 54:278-279.

_____. 1989. Dental attrition and enamel hypoplasia among late Pleistocene immature hominids from western Europe. *American Jour. Phys. Anthropol.* 78:303-304.

Smirnov, Y. 1989. Intentional human burial: Middle Palaeolithic (last glaciation) beginnings. *Jour. World Prehist.* 3:199-233.

Smith, B. H. 1991. Dental development and the evolution of life history in Hominidae. *American Jour. Phys. Anthropol.* 86:157-174.

_____. 1993. The physiological age of KNM-WT17000. Pages 195-220 *in* A. Walker & R. Leakey, eds., *The Nariokotome* Homo erectus *Skeleton*. Harvard University Press, Cambridge, Massachusetts.

Smith, F. H. 1976. *The Neandertal Remains from Krapina: A Descriptive and Comparative Study*. Department of Anthropology, Report No. 15, University of Tennessee, Knoxville, Tennessee. 359 pp.

_____. 1992. The role of continuity in modern human origins. Pages 145-156 *in* G. Bräuer & F. H. Smith, eds., *Continuity Or Replacement: Controversies in* Homo sapiens *Evolution*. A. A. Balkema, Rotterdam, Netherlands.

_____. 1994. Samples, species and speculations in the study of modern human origins. Pages 227-249 *in* M. H. Nitecki & D. V. Nitecki, eds., *Origins of Anatomically Modern Humans*. Plenum Press, New York.

_____, & J. Ahern. 1994. Additional cranial remains from Vindija Cave, Croatia. *American Jour. Phys. Anthropol.* 93:275-280.

Soffer, O. 1985. Patterns of intensification as seen from the Upper Paleolithic of the central Russian plain. Pages 235-270 *in* T. C. Price & J. A. Brown, eds., *Prehistoric Hunters and Gatherers*. Academic Press, Orlando, Florida.

_____. 1992. Social transformation at the Middle to Upper Palaeolithic transition: The implications of the European record. Pages 247-259 *in* G. Bräuer & F. H. Smith, eds., *Continuity or Replacement: Controversies in* Homo sapiens *Evolution*. A. A. Balkema, Rotterdam, The Netherlands.

Solecki, R. 1971. *Shanidar: The First Flower People.* Alfred A. Knopf, New York. 290 pp.

_____. 1975. Shanidar IV: A Neanderthal flower burial in northern Iraq. *Science* 190:880-881.

Spiess, A. E. 1979. *Reindeer and Caribou Hunters: An Archaeological Study.* Academic Press, New York. 312 pp.

Stahl, A. B. 1984. Hominid dietary selection before fire. *Curr. Anthropol.* 25:151-168.

Stanford, D. 1979. Bison-kill by ice-age hunters. *Natl. Geogr. Mag.* 155(1):114-121.

Stiner, M. C. 1991. The faunal remains at Grotta Guattari: A taphonomic perspective. *Curr. Anthropol.* 32:103-117.

_____. 1994. *Honor Among Thieves: A Zooarchaeological Study of Neandertal Ecology.* Princeton University Press, Princeton, New Jersey. 422 pp.

Straus, L. G. 1994. Upper Paleolithic origins and radiocarbon calibration: More new evidence from Spain. *Evol. Anthropol.* 2:195-198.

_____, J. Bischoff, & E. Carbonell. 1993. A review of the Middle to Upper Paleolithic transition in Iberia. *Préhistoire Européene* 3:11-27.

Stringer, C. 1992. Replacement, continuity and the origin of modern humans. Pages 9-24 *in* G. Bräuer & F. H. Smith, eds., *Continuity or Replacement: Controversies in* Homo sapiens *Evolution.* A. A. Balkema, Rotterdam, Netherlands.

_____. 1994. Out of Africa—a personal history. Pages 149-172 *in* M. H. Nitecki & D. V. Nitecki, eds., *Origins of Anatomically Modern Humans.* Plenum Press, New York.

_____, M. C. Dean, & R. D. Martin. 1990. A comparative study of cranial and dental developments within a recent British sample and among Neandertals. Pages 115-152 *in* C. J. DeRousseau, ed., *Primate Life History and Evolution.* Wiley-Liss, New York.

Susman, R. L. 1988. Hand of *Paranthropus robustus* from Member 1, Swartkrans: Fossil evidence for tool behavior. *Science* 240:781-784.

_____, J. T. Stern, & W. L. Jungers. 1984. Arboreality and bipedality in the Hadar hominids. *Folia Primatol.* 43:113-156.

Sutcliffe, A. J. 1970. The spotted hyaena: Crusher, gnawer, digester, and collector of bones. *Nature* 227:1110-1113.

Swisher, C. C. III, G. H. Curtis, T. Jacob, A. G. Getty, A. Suprijo, & Widiasmoro. 1994. Age of the earliest known hominids in Java, Indonesia. *Science* 263:1118-1121.

Tappen, M. 1992. *Taphonomy of a Central African Savanna: Natural Bone Deposition in the Parc National des Virunga.* Ph.D. Dissertation, Harvard University, Cambridge, Massachusetts. 300 pp.

Taylor, C. R., & V. J. Rowntree. 1973. Running on two or four legs: Which consumes more energy. *Science* 179:186-187.

Tattersall, I. 1986. Species recognition in human paleontology. *Jour. Hum. Evol.* 15:165-176.

Tchernov, E. 1987. The age of the Ubeidiya Formation: An early Pleistocene hominid site in the Jordan Valley, Israel. *Israel Jour. Earth Sci.* 36:3-30.

_____. 1988. The paleobiogeographical history of the southern Levant. Pages 159-250 *in* Y. Yom-Tov & E. Tchernov, eds., *The Zoogeography of Israel: The Distribution and Abundance at a Geological Crossroad.* Monographiae Biologicae, Dordrecht, The Netherlands.

Tobias, P. V. 1991. *Olduvai Gorge, Volume 4: The Skulls, Endocasts and Teeth of* Homo habilis. Cambridge University Press, Cambridge, UK. 921 pp.

Tooby, J., & I. DeVore. 1987. The reconstruction of hominid behavioral evolution through strategic modeling. Pages 183-237 *in* W. Kinzey, ed., *The Evolution of Human Behavior: Primate Models.* SUNY Press, Albany, New York.

Toth, N. 1987. Behavioral inferences from early stone age artifact assemblages: An experimental model. *Jour. Hum. Evol.* 16:763-787.

_____, & K. D. Schick. 1986. The first million years: The archaeology of protohuman culture.

Pages 1-96 *in* M. B. Schiffer, ed., *Advances in Archaeological Method and Theory*, volume 9. Academic Press, Orlando, Florida.

_____. 1993. Early stone industries and inferences regarding language and cognition. Pages 346-362 *in* K. R. Gibson & T. Ingold, eds., *Tools, Language and Cognition in Human Evolution*. Cambridge University Press, Cambridge, UK.

Trinkaus, E. 1981. Neanderthal limb proportions and cold adaptation. Pages 187-224 *in* C. B. Stringer, ed., *Aspects of Human Evolution*. Taylor & Francis, London, UK.

_____. 1983a. *The Shanidar Neandertals*. Academic Press, New York. 502 pp.

_____. 1983b. Neandertal postcrania and the adaptive shift to modern humans. Pages 165-200 *in* E. Trinkaus, ed., *The Mousterian Legacy*. B.A.R. International Series 164, Oxford.

_____. 1985. Cannibalism and burial at Krapina. *Jour. Hum. Evol.* 14:203-216.

_____. 1986. The Neandertals and modern humans origins. *Annu. Rev. Anthropol.* 15:193-218.

_____. 1989. The Upper Pleistocene transition. Pages 42-66 *in* E. Trinkaus, ed., *The Emergence of Modern Humans: Biocultural Adaptations in the Later Pleistocene*. Cambridge University Press, Cambridge, UK.

_____. 1993. Neandertal mortality patterns: Stress and/or sampling bias. Paper [oral] presented at the annual meeting of the Palaeoanthropology Society, April 13-14, Toronto, Canada.

_____, S. Churchill, & C. B. Ruff. 1994. Postcranial robusticity in *Homo*: Humeral bilateral asymmetry and bone plasticity. *American Jour. Phys. Anthropol.* 93:1-34.

_____, & R. L. Tompkins. 1990. The Neandertal life cycle: The possibility, probability, and perceptibility of contrasts with recent humans. Pages 153-180 *in* C. J. DeRousseau, ed., *Primate Life History and Evolution*. Wiley-Liss, New York.

Troeng, J. 1993. *Worldwide Chronology of Fifty-three Prehistoric Innovations*. Acta Archaeol. Lundensia Series In 8° No. 21. Almqvist & Wicksell International, Stockholm, Sweden. 311 pp.

Tuffreau, A., & J. Somme. 1988. Le gisement paléolithique moyen de Biache-St-Vaast (Pas de Calais). Vol. 1: Stratigraphie, environment, études archéologiques (1ère partie). *Mem. Soc. Préhist. Française* 21:1-307.

Turner, A., & A. Chamberlain. 1989. Speciation, morphological change, and the status of African *Homo erectus*. *Jour. Hum. Evol.* 18:115-130.

Turner, C. G. II. 1983. Dental evidence for the peopling of the Americas. Pages 147-157 *in* R. Shutler, ed., *Early Man in the New World*. Sage Publications, Beverly Hills, California.

_____. 1986. Dentochronological separation estimates for Pacific rim populations. *Science* 232:1140-1142.

_____. 1989. Teeth and prehistory in Asia. *Sci. American* 260(2):88-94.

Tuross, N., & M. Fogel. 1993. Stable isotope analysis and subsistence patterns at the Sully site. Pages 283-290 *in* D. Owsley & R. Jantz, eds., *Skeletal Biology in the Great Plains*. Smithsonian Institution Press, Washington, DC.

Valladas, H., J. L. Reyss, J. Joron, G. Valladas, O. Bar-Yosef, & B. Vandermeersch. 1988. Thermoluminescence dating of Mousterian "proto-Cro-Magnon" remains from Israel, and the origin of modern man. *Nature* 331:614-616.

Vandermeersch, B. 1981. *Les Hmmes Fossiles de Qafzeh (Israel)*. Éditions du CNRS, Paris, France. 319 pp.

Vandiver, P. B., O. Soffer, B. Klima, & J. Svoboda. 1989. The origins of ceramic technology at Dolni Vestonice, Czechoslovakia. *Science* 246:1002-1008.

Van Noten, F. 1977. Excavations at Matupi Cave. *Antiquity* 51:35-40.

Vaughan, P. 1985. *Use-wear Analysis of Flaked Stone Tools*. University of Arizona Press, Tucson, Arizona. 204 pp.

Vermeersch, P. 1990. Palaeolithic chert exploitation in the limestone strata of the Egyptian Nile Valley. *Afr. Archaeol. Rev.* 8:77-102.

Villa, P. 1990. Torralba and Aridos: Elephant exploitation in Middle Pleistocene Spain. *Jour. Hum. Evol.* 19:299-309.

Volman , T. P. 1984. Early prehistory of southern Africa. Pages 169-220 *in* R. G. Klein, ed., *Southern African Prehistory and Palaeoenvironments*. A. A. Balkema, Rotterdam, The Netherlands.

Wadley, L. 1993. The Pleistocene Later Stone Age south of the Limpopo River. *Jour. World Prehist.* 7:243-296.

Walker, A. 1993. Perspective on the Nariokotome discovery. Pages 411-430 *in* A. Walker & R. Leakey, eds., *The Nariokotome* Homo erectus *Skeleton*. Harvard University Press, Cambridge, Massachusetts.

_____ , & R. Leakey, eds. 1993. *The Nariokotome* Homo erectus *Skeleton*. Harvard University Press, Cambridge, Massachusetts. 457 pp.

Wendorf, F. 1968. Site 117: A Nubian final Paleolithic graveyard near Jebel Sahaba, Sudan. Pages 954-995 *in* F. Wendorf, ed., *The Prehistory of Nubia*, volume II. Southern Methodist University Press, Dallas, Texas.

_____ , R. Schild, & A. Close. 1993. *Egypt During the Last Interglacial: The Middle Palaeolithic of Bir Tarfawi and Bir Sahara East*. Plenum Press, New York. 596 pp.

Wendt, W. E. 1972. Preliminary report on an archaeological research programme in South West Africa. *Cimbebasia* B2:1-61.

Wheeler, P. E. 1990. The significance of selective brain cooling in hominids. *Jour. Hum. Evol.* 19:321-322.

_____ . 1991a. The thermoregulatory advantages of hominid bipedalism in open equatorial environments: The contribution of increased convective heat loss and cutaneous evaporative cooling. *Jour. Hum. Evol.* 21:107-115.

_____ . 1991b. The influence of bipedalism on the energy and water budgets of early hominids. *Jour. Hum. Evol.* 21:117-136.

_____ . 1992a. The thermoregulatory advantages of large body size for hominids foraging in savannah environments. *Jour. Hum. Evol.* 23:351-362.

_____ . 1992b. The influence of the loss of functional body hair on the water budgets of early hominids *Jour. Hum. Evol.* 23:379-388.

White, R. 1990. Production complexity and standardization in early Aurignacian bead and pendant manufacture: Evolutionary implications. Pages 366-390 *in* P. Mellars & C. B. Stringer, eds., *The Human Revolution: Behavioural and Biological Perspectives on the Origins of Modern Humans*. Edinburgh University Press, Edinburgh, UK.

White, T. D. 1986. Cut marks on the Bodo cranium: A case of prehistoric defleshing. *American Jour. Phys. Anthropol.* 69:503-509.

Wiessner, P. 1983. Style and information in Kalahari San projectile points. *American Antiquity* 48:253-276.

_____ . 1984. Reconsidering the behavioral basis for style: A case study among the Kalahari San. *Jour. Anthropol. Archaeol.* 3:190-234.

Wilmsen, E. N. 1974. *Lindenmeier: A Pleistocene Hunting Society*. Harper & Row, New York. 126 pp.

Willoughby, P. J. 1985. Spheroids and battered stones in the African Early Stone Age. *World Archaeol.* 17:44-60.

Wolpoff, M. H., F. H. Smith, M. Malez, J. Radovcic, & D. Rukavina. 1981. Upper Pleistocene human remains from Vindija Cave, Croatia, Yugoslavia. *American Jour. Phys. Anthropol.* 54:499-545.

Woo, J. 1966. The skull of Lantian man. *Curr. Anthropol.* 7:83-86.

Wood, B. 1984. The origin of *Homo erectus. Cour. Forschungsinst. Senckenb.* 69:99-111.

_____. 1991. *Koobi Fora Research Project, Volume 4: Hominid Cranial Remains.* Clarendon Press, Oxford, UK. 466 pp.

_____. 1992a. Origin and evolution of the genus *Homo. Nature* 355:783-790.

_____. 1992b. Early hominid species and speciation. *Jour. Hum. Evol.* 22:351-365.

_____. 1993. Early *Homo*: How many species? Pages 485-522 *in* W. H. Kimbel & L. B. Martin, eds., *Species, Species Concepts, and Primate Evolution.* Plenum Press, New York.

Wu, R., & J. W. Olsen, eds. 1985. *Palaeoanthropology and Palaeolithic Archaeology in the People's Republic of China.* Academic Press, Orlando, Florida. 293 pp.

Yellen, J. E. 1991. Small mammals: Post-discard patterning of !Kung San faunal remains. *Jour. Anthropol. Archaeol.* 10:152-192.

_____, A. S. Brooks, E. Cornelissen, M. Mehlman, & K. Stewart. 1995. A Middle Stone Age worked bone industry from Katanda, Semliki Valley, Zaire. *Science* 268:553-556.

Zhou, M. Z., & C. K. Ho. 1990. History of the dating of *Homo erectus* at Zhoukoudian. *Geol. Soc. America Spec. Pap.* 242:69-74.

Molecular Anthropology in Retrospect and Prospect

Jonathan Marks

Department of Anthropology
Yale University
New Haven, CT 06511

The application of molecular genetic data to anthropological questions has a long history. These data, like any data, only make sense as they are integrated into the existing corpus of data and theory. Our cultural prejudice about the primacy of heredity in human affairs has sometimes allowed studies based on genetic data to gain more credibility than they merit from methodological, analytical, and theoretical standpoints. Ultimately molecular data augment, but do not supersede, more traditional methods of anthropological inquiry.

I think we are all in fundamental agreement about the utility of molecular data in augmenting the more traditional modes of scientific inference about human origins. Genetic data provide for us an independent test of phylogenetic hypotheses; independent, that is, of the morphological, anatomical data with which macrobiologists generally work and on the basis of which they have been eminently successful in reconstructing the history of life.

Occasionally, however, there is a specific phylogenetic problem that appears to be intractable. Perhaps the species have adapted to radically different environments, concealing much of their shared ancestry; or perhaps several species all appear to be equally different from one another. In such cases, genetic data can be helpful, providing a suite of characteristics — the nucleotides that compose DNA, or some estimator of them — that do not directly interact with the environment themselves, and therefore will be less subject to convergent evolution than anatomical traits.

Often, however, we expect more from genetic data in the arena of anthropological systematics. It is not simply that we have another set of data that transcends the main problems in the data we are accustomed to. Rather, there is the expectation that in the genetic data we have something truer and more scientifically valid, something encoding actual history so purely that all you have to do is put on the right pair of glasses and read it.

I will argue in this paper that the right pair of glasses are rose colored, and that our attitude toward genetic studies in anthropology is far more deeply rooted in our cultural values than in any significant aspects of genetic data themselves. It lies in the Euro-American tradition that "blood will tell," which is still with us socially and scientifically. Science paperbacks, and some scientists themselves, now assert that deriving the sequence of the DNA of a single composite human cell — a genome — will cure cancer (Dulbecco 1986) and social problems like homelessness (Koshland 1989) and reveal ultimate insights into the human condition (Watson 1990). Indeed, the goal of sequencing a single human genome has become largely synonymous with the research enterprise of human molecular genetics (Olson 1993).

Now, I would not use genetic methodologies in my own research if I did not think them worthwhile. The criticism I will raise — the set of cultural values of which I speak — is very specifically in the taking of genetic data uncritically as authoritative. Introspective anthropologists have shown that we generally project our cultural values and expectations onto the data of paleoanthropology and primate behavior, in order to come up with an explanatory narrative of human evolution. But somehow, genetic data are above all that: they involve fancy equipment and computers, and we have come to expect that they tell a story almost independently of such mundane considerations as the assumptions, data analysis, experimental design, or theoretical basis for interpreting the results.

Yet historically there are myriad examples of how genetic work has been loaded with cultural values and has guided us to wrong-headed conclusions (Nelkin & Lindee 1995). In 1924, for example, every textbook of genetics discussed the topic of eugenics favorably, and every geneticist of note in America advocated that program (Marks 1993b). This was the program that held most of the world's populations to be constitutionally degraded, possessing a recessive allele for feeble-mindedness. Therefore, the solution to urban social problems was to discourage immigrants from breeding and to prevent more immigrants from entering. It provided the scientific validation for the Johnson Immigration Restriction Act of 1924.

Nowadays, we have the disputes over DNA fingerprinting, which had been used to bully juries intellectually until 1989, when a couple of clever lawyers actually had other geneticists examine the raw genetic data constituting the scientific validation for the state's case. They found the controls to be inadequate, the experiments shabbily conducted, and they began to raise basic questions about the validity of conclusions drawn from such work, questions that have not yet been satisfactorily resolved (Billings 1992).

From the standpoint of human origins, however, the controversies surrounding DNA fingerprinting or an egregious social program like eugenics are about heredity *per se*, but not really about systematics. I introduce them to illustrate the fact that genetic data are fallible; or more specifically, interpretations of genetic data are fallible. This has direct relevance for anthropological systematics.

Genetics in Racial Anthropology

Physical anthropology in pre-1960 days centered around the question of identifying the single-digit number of basic forms into which the human species could be subdivided. These forms, of course, were called races, and their differences were thought to be the source of the bulk of the hereditary variation in *Homo sapiens*. Fieldwork led anthropologists to the conclusion that races were themselves exceedingly heterogeneous; studies of immigrants showed that much of the anatomy that had been considered fundamentally racial, such as head shape and other aspects of body form, was in fact strongly influenced by the environment.

The alarm call in racial studies was answered by the study of blood. Blood, of course, is a dominant metaphor for heredity, and in this case one could study the blood groups of different peoples and compare them over one tiny portion of their hereditary makeup.

Now, let us cheat a bit. There are three major alleles, A, B, and O, and pretty much all populations have all three alleles. Geneticists call this a polymorphism (Table 1). In fact, that is paradigmatic for the study of human genetic variation: nearly all populations have nearly all alleles. Virtually everywhere, O is the most common allele. A is very common in all populations, except among Native Americans. B is uncommon in all populations, but more common among Asians than among most others. Some Native American groups seem to have lost both the A and B alleles, being almost totally O. The differences we encounter among ABO frequencies across populations are intergrading, not discrete, and it seems unreasonable to expect that they would shed much light on the discrete differences that anthropologists wished to identify as being racial.

Nevertheless they did, according to geneticists of the 1920s and 1930s. The initial ABO frequency data collected by Hirschfeld and Hirschfeld (1919) during World War I managed to divide populations into three categories: European, Asio-African, and Intermediate. In other words, it partitioned the human species pretty nicely into "white" and "other." Since the ABO blood group says nothing of the sort nowadays,

TABLE 1. ABO Allele Frequencies from Representative Populations.
Taken from Mourant *et al.* (1976)

Aboriginal Population		*A*	*B*	*O*
America	Chippewa	.06	.00	.94
	Kwakiutl	.10	.00	.90
Europe	Denmark	.27	.08	.66
	Bulgaria	.31	.12	.56
	Ukraine	.27	.16	.57
Asia	Kazakhstan	.25	.27	.48
	Pakistan	.20	.25	.55
	Japan	.29	.16	.54
Oceania	Australia	.18	.04	.78
Africa	Efe Pygmies	.26	.21	.53
	Angola	.16	.11	.72
	Sierra Leone.	.16	.15	.69

it stands to reason that those conclusions tell us more about the researchers' mindset than it does about genetics.

But that was not all, for there was prehistory to reconstruct. The presence of three diagnosably different kinds of blood substances (A, B, and O) made it, in the words of those researchers, "very difficult to imagine one single place of origin for the human race" (Hirschfeld & Hirschfeld 1919:679). Consequently, they proposed an ancestral "O" human species, subsequently invaded by "two different biochemical races which arose in different places" (*ibid*.). In other words, polymorphism and heterozygosity were the result exclusively of racial invasions.

Encountering almost wholly type-O blood in the New World led genetic serologists a few years later to the conclusion that Native Americans diverged from the rest of humanity early on, before the invasions of the A and B people in Europe, Asia, and Africa and the differentiation of those populations from one another (Coca & Diebert 1923). That this interpretation agreed with virtually nothing from the field of anthropology did not seem to matter.

In 1925 and 1926 the blood group serologists, apparently dissatisfied with the "white/other" classification, used ABO to divide humanity into seven "types": European, Intermediate, Hunan, Indo-Manchurian, Africo-Malaysian, Pacific-American, and Australian (Ottenberg 1925; Snyder 1926). These were at some variance from the groups anthropologists tended to see when they divided up the human species, but this was genetic data, and there seemed no good reason even to try to reconcile it to anthropology.

Actually, however, there was no real division of these types strictly on the basis of their ABO allele frequencies. What the researchers had done was just basically to divide the world's populations into large para-continental groups, impose the ABO data upon them, and describe the results. Consequently, several populations assigned to one "type" actually had ABO frequencies that fell within the ranges of other human "types." Conversely, in some cases diverse people happened to have too similar a distribution of alleles. This produced a number of inconsistencies. For example, the people of Senegal, Vietnam, and New Guinea ended up together. Likewise did the people of Poland and China. America's leading physical anthropologist, Harvard's Earnest Hooton, whose interest lay in isolating pure racial (pheno)types, felt that any method that put together such a motley group of people was basically worthless (1931:490).

By 1930, serologist Laurence Snyder had abandoned the seven race-type system, but still argued "forcibly [for] the value of the blood groups as additional criteria of race-classification" (1930:132). Snyder now had the peoples of the world carved up into 25 (unnamed) clusters, based on different criteria than anthropologists used, but harmonious to some extent and disharmonious to some extent. His prediction that "in the future no anthropologic study will be complete without a knowledge of the blood group proportions under discussion" (*ibid*.) could be considered optimistic, given the specious reasoning and conclusions that had accompanied its use thus far. Indeed, the fact that apes have ABO alleles, which strongly suggests that the polymorphism is ancestral to the origin of the human species, was demonstrated in 1925, but the population dynamics of a primitively polymorphic system does not seem to have been invoked as an explanation for the ABO distribution in humans before William Boyd (1940).

Where there was concordance between genetic and phenotypic analyses, obviously no difficulties arose. Where there was discordance, however, the assumption that results derived from the blood must be fundamentally more reliable, no matter how they were generated or how poorly understood they were, remained a justification for using the serological data to carve up the human species. It took decades of mulling over these data to appreciate the basic nature of the patterns being encountered in the human species. And the basic pattern was what only Hooton recognized: if you were interested in establishing discrete phenotypic groups of people, the genetic data were pretty much irrelevant.

Ultimately, the major original contribution of serological data in racial anthropology was to promote the Basque people of the Pyrenees of Spain to a separate race, equivalent to "Africans" or "American Indians" (Boyd 1950:268). Of course, it's not as if they had green skin and square heads; they simply speak a strange language and, more importantly, have allele frequencies somewhat different from their neighbors. And that is enough, if you believe in the existence of discrete races and in the power of genetics — just a little bit of genetics — to reveal them.

So the study of blood group genetics, cited since the early 1970s as undermining the concept of race, was consistently interpreted within it, and certainly as supporting it. The races revealed were sometimes concordant with traditional anthropological ones (especially when the genetic data were largely imposed on geography); when they were discordant, the genetic data were assumed to be simply better. There were no good scientific reasons for this, just good old cultural ones.

Probably the most interesting claim for the hereditary study of our species was published in the *American Journal of Physical Anthropology* in 1927 by a Russian named Manoilov, who reported that a series of simple chemicals added to a sample could distinguish Russian blood from Jewish blood. Following this, Poliakowa (1927) reported that Manoilov's test permitted her to distinguish among the bloods of various Euro-Asian countries. The test turned the blood of Russians reddish, of Jews blue-greenish, and of various other peoples various other colors. Again, America's leading student of human diversity as race, Earnest Hooton, found this claim difficult to swallow (1931:491).

In other words, this study of "blood" — again, a dominant metaphor for "heredity" — was producing conclusions at considerable variance with what anthropologists already knew. Races, whatever they were, were certainly not the same as nations, and it was hard to believe that they would each have diagnosably different blood structures. The notable aspect of this critique is that it is not based on any methodological flaw in the work, but simply on the conclusions, which could only have been a reflection of the investigators' prejudices or expectations. The work of the Russian hematologists was rejectable because in their ignorance they failed to appreciate that known patterns of human variation invalidated their work *a priori*; in their arrogance, they thought they were going to revolutionize anthropology with it. The work was scientific, it involved strict hereditary factors, it was reviewed and published, and it came to foolish conclusions that were dismissible independently of the allure of the hereditary substance they studied.

In a follow-up publication also in the *American Journal of Physical Anthropology*, Manoilov (1929) reported the ability to distinguish the blood of men from women, and discussed a colleague whose application of these techniques could distinguish the blood of homosexuals from heterosexuals. And it worked just as well on plants, in spite of the biological difficulty posed by extracting blood from them. I would suggest that if these workers had made identically idiotic claims from measuring the width of the shoulder blade, their manuscripts would never have made it past the secretary's desk. The fact that they were studying blood, the fact that they were studying deep heredity, gave their work greater power and validation than it actually merited.

Manoilov's work was far from obscure, for it was discussed at the highest levels in the genetics community — by T. H. Morgan and C. B. Davenport — and was the subject of several publications in *Science* and the *Proceedings of the National Academy of Sciences*. It also found its way into genetics textbooks of the era (Schull 1931; Sinnott & Dunn 1932). Interest among the cognoscenti had begun to wane after a biochemist reported privately that Manoilov's blood test did not actually even require blood; it worked just as well on urine (K. George Falk to C. B. Davenport, 9 December 1926, C. B. Davenport Papers, American Philosophical Society). To this day we do not know what the test was actually testing — quite possibly a figment of geneticists' imaginations, yielding essentialized differences among groups presumed to be essentially different.

The Trichotomy: Molecular Anthropology's Gordian Knot

The undisputed triumph of molecular work in anthropological systematics was Morris Goodman's demonstration (1962) that humans and chimpanzees and gorillas were more closely related to each other than any was to an orangutan. Not only that, but genetically — if no other way — humans, chimpanzees, and gorillas were virtually indistinguishable from one another. They appeared to form a genetic "trichotomy." (The de-hominization of the fossil taxon *Ramapithecus* (Sarich & Wilson 1967), often hailed as a triumph of molecular systematics in anthropology (*e.g.*, Lewin 1987), was actually a dispute over rates of molecular evolution, and its implications for systematics were thus indirect. Nevertheless, for those with a short memory, it seemed to bring the score to Molecules 1, Morphology 0.)

In the 1980s, however, it emerged as threatening that the ostensibly potent molecular data were apparently unable to resolve this three-way split into two sequential two-way splits. An extraordinary amount of effort has been thrown into the question of genetically resolving this trichotomy — as if a three-way split were impossible, and as if genetic data, like the Delphic Oracle, must be capable of answering all questions put to them.

The gross comparison of ape DNA, by DNA hybridization, shows a three-way

TABLE 2. DNA hybridization values from two studies, mean distances and standard deviations. Column (A) is the numbers given by Sibley *et al.* (1990), without the impermissible data alterations used, and not reported, in their prior publications. Column (B) is calculated from the Appendix to Caccone & Powell (1989), without the alteration based on the indefensibly precise measurement of DNA fragment length. These are the most comparable sets of numbers; see Marks (1991).

	Unaltered ΔT_m Values	
	(A)	(B)
Human-Chimpanzee	$1.4 \pm .8$	2.4 ± 1.1
Chimpanzee-Gorilla	$1.7 \pm .4$	$3.0 \pm .6$
Human-Gorilla	$1.8 \pm .8$	4.1 ± 1.1

split, with large error margins (Table 2). This technique, it turns out, only resolves the trichotomy when you tinker with the numbers; for example, by substituting controls across experiments, or moving correlated points into a regression line (Sibley *et al.* 1990), or precisely changing the measured values on the basis of a DNA smear from which the DNA fragment length is precisely extracted to the nearest single nucleotide (Caccone & Powell 1989), which users of this technique admit they did (Marks 1991). The fact is you could resolve anything with any set of data if taking such liberties were legitimate. It is not, and general interest and use of this technique has precipitously declined on that basis. That a few zealots still defend it (Lowenstein 1993; Ruvolo 1995) actually speaks more to sociology-of-science than to molecular anthropology. At best, the work represented the Manoilov blood test of the 1980s.

DNA sequencing studies fall fairly neatly into two categories. The first comprises those in which the split is too close to call, and its resolution depends heavily on the method of outgroup rooting and clustering technique. The second category comprises those in which a specific pairwise linkage is argued to be favored, contradicting other data sets in which a different pairwise linkage is favored (Marks 1992).

This raises fascinating questions about the epistemology of molecular evolution. For example, although three data sets of mitochondrial DNA fall into the first category (Ferris *et al.* 1980; Brown *et al.* 1982; Hixson & Brown 1986), a fourth was claimed to have resolved the trichotomy into human-chimpanzee (Horai *et al.* 1992). The reasoning is simple: these 4900 bases yield approximately 180 phylogenetically informative nucleotide sites (*i.e.*, those in which two character states are found across human, chimpanzee, and gorilla, and one of them is also in the outgroup, the orangutan). Approximately 85 of these link human-chimpanzee, 55 link chimpanzee-gorilla, and 40 link human-gorilla. Since the favored pairing here is human-chimpanzee, the link is taken as proved.

But evolutionary history is not like a Democratic primary in which the candidate with the plurality of votes wins. For if those 85 bits of data are giving the "right" answer, then the 95 other bits of data (55 + 40) in that same set are giving the "wrong" answer. The argument for human-chimpanzee, then, boils down to accepting 47% of the informative data, and rejecting 53%. Seen this way, it does not constitute a particularly strong scientific argument in favor of resolution of the trichotomy.

The key recognition here undermining the claim of "resolution" lies in considering the efficiency of this molecular evolutionary system. To judge the phylogeny as "resolved" implies that this system is inefficient, or un-parsimonious, enough to generate 53% "wrong" answers, if we assume that the 47% showing human-chimpanzee are "right." But paradoxically, it also implies that this system is so efficient that it could not be generating 70% "wrong" answers and only 30% "right" answers (*i.e.*, if chimpanzee-gorilla were "right"). In other words, the assumption is that this system is maximally parsimonious, although it is obviously not very parsimonious at all.

In my opinion the best interpretation is actually neither of these. Rather, it is that these data reinforce the trichotomous nature of the relations among these three lineages. If it were indeed the case that the late Miocene stem lineage effectively split simultaneously into three lines, we would expect those 180 informative sites to distribute themselves as about 60-60-60 in favor of each of the three pairwise linkages, and thus only 25 of them, or 14%, are "wrong."

The second category, datasets pointing one way and contradicted by others pointing a different way, represents much of the DNA data from the nucleus (Marks 1992). Most of our data here come from the beta-globin cluster, of which bits and pieces have told various stories at various times (Miyamoto & Goodman 1990), although generally tending to point to human-chimpanzee (Bailey *et al.* 1991). The gene for a skin protein called involucrin seems to link chimpanzee and gorilla (Djian & Green 1989). The immunoglobulin genes yield all three permutations (Ueda *et al.* 1985; Ueda *et al.* 1989; Kawamura *et al.* 1991). Best of all, the little bit of DNA where the X and Y chromosomes pair up, known as the pseudoautosomal region, slightly favor chimpanzee-gorilla with the X, and chimpanzee-human with the Y (Ellis *et al.* 1990).

The trouble here in gene-land is actually very fundamental. It is the absence of population genetics, of a consideration for the processes of microevolution that might impact upon the clarity of discernible macroevolutionary patterns (Rogers 1993) — not terribly unlike the early racial serology.

In one of the most important, but unheralded, studies of recent years in molecular anthropology, Ruano *et al.* (1992) actually looked at the extent of within-species diversity in the apes, in relation to the phylogenetic information extractable from the DNA sequence they were studying. They looked at a stretch of a couple of hundred nucleotides from the homeobox cluster of chromosome 17. Finding it to be absolutely invariant across a diverse sample of humans from all over the world, they then studied 16 homologous chimpanzee sequences and found two alleles, and twenty gorilla sequences and found four alleles, one of which was very divergent (Figure 1). Their important conclusion was that, despite a single DNA base difference superficially linking the human to the two chimpanzee sequences, the amount of polymorphism present, reflecting the breadth of the gene pool, swamps any attempt to extract a reliable phylogenetic inference here.

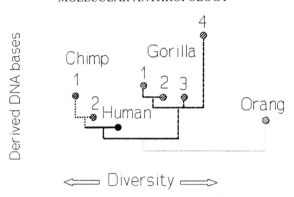

FIGURE 1. Extensive diversity in African ape DNA, drawn from very heterogeneous gene pools and undermining the prospects of "resolving the trichotomy" from single representatives of each species.

There are two important theoretical reasons why polymorphism is necessary to sample in a problem like this. The first involves the simple discordance between the phylogenetic history of three species and of three DNA sequences when the three DNA sequences are taken to represent the three taxa (Figure 2). If the dark allele gives rise to the intermediate allele, and the intermediate allele to the light allele, and all three co-exist in the ancestral species, they may segregate into descendant taxa in a manner that reflects not the sequence of speciation events (or the species tree), but only the descent of the DNA sequences from one another (or the gene tree). This can hopefully be overcome by sequencing many unlinked loci, which would presumably not all be segregating in such a fashion (Pamilo & Nei 1988).

The second problem is more formidable and stems from the fact that the budding

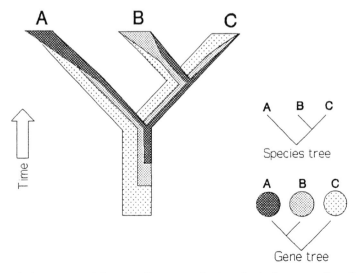

FIGURE 2. A gene tree might not replicate a species tree, due to the segregation of ancestral polymorphism. See text for explanation. (After Marks 1992)

of a new species from an old, speciation by the founder effect, which is probably very common, results in paraphyly in the parent species. Paraphyly is a relationship wherein all members of a biological group are not one another's closest relatives, because one or more that should be in there have been excluded.

In Figure 3, we have a cosmopolitan species, A, which varies genetically over a geographical gradient, as any species in nature will. From some part of this species, a new species, B, forms. B forms from members not just of species A, but from a particular segment of species A, and then obtains its own biological identity. Species A is now paraphyletic because, regardless of who can mate with whom, which is usually what we mean by species, species B is genetically more closely related to one part of species A than two parts of species A are to one another. Now, species A and species C are still one species. Species C forms some time later, from another segment of species A. When we draw the relationships based on this simple history, we would therefore draw it with A and C sharing a unique chunk of biological history, not shared by B. Thus, A and C are closest relatives.

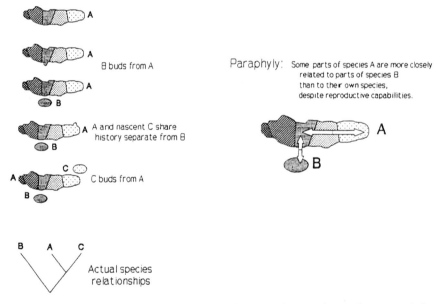

FIGURE 3. Founder effect speciation in a variable ancestral population produces paraphyly.

But if we try to reconstruct that history from a single piece of DNA from each species — no matter how much DNA — we run into a problem (Figure 4). If the representative of species A that we choose is from the left, we may find B and C to be closest relatives; if it is from the center, we may find A and B to be closest relatives. We would really have to be lucky and choose a specimen from the right to come up with the historically correct answer. In other words, you need several specimens of species A in order to have any kind of a fair chance of reconstructing the biological history of this group, and the chances of doing so will be affected by variables like

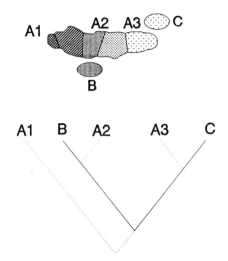

FIGURE 4. Paraphyly produces phylogenetic ambiguities.

the time between the two divergences and the extent of gene flow within the parental species.

We can add another fairly simple twist to this example. Say that species C is very successful, out-competes part of species A, and expands its range at the expense of species A (Figure 5). This could quite simply wipe out your chance of recovering the correct relationships altogether. All the more reason why I would expect molecular studies of the apes to understate the strength of their claims to have resolved the trichotomy in the absence of information on intra-specific variation, but generally they do not. There is, after all, something of a lack of humility often characterizing molecular research generally. It feeds off our cultural values of what we think is important and fundamental and revelatory — it is a glimpse of deep heredity, although sometimes in the absence of a sound theoretical framework for understanding or interpreting it.

It is of some interest in this regard that the sets of genetic data which have actually attempted to sample intraspecific variation generally tend to link chimpanzee to gorilla, in harmony with traditional interpretations of the anatomy. These are generally qualitative studies as well: the involucrin gene sequence (Djian & Green 1990); the dark bands at the chromosome tips of chimpanzees and gorillas (Marks 1993a); as well as other DNA sequences (Dangel *et al.* 1995; Livak *et al.* 1995; Meyer *et al.* 1995). One quantitative study examining intraspecific variation and linking chimpanzee to human has also been reported (Ruvolo 1994), but its proper interpretation is unclear (Marks 1995a; Green & Djian 1995; Rogers & Comuzzie 1995). The strongest molecular evidence in support of human-chimpanzee is from the hemoglo-

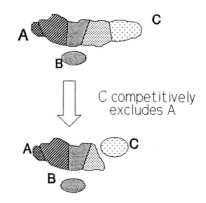

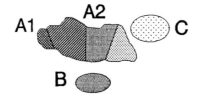

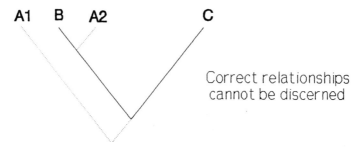

FIGURE 5. Under some simple circumstances phylogeny may not be accurately recoverable. See text for discussion.

bin genes on chromosome 11, which is all-too-familiar to anthropologists as the site of strong selection for genetic diversity leading to malaria resistance — although no polymorphism has been sampled in the phylogenetic research here.

Paradoxically, in this arena the studies making the most extravagant claims often emerge to have been based on the weakest data. One widely-publicized study claimed "resolution" of the trichotomy based on 692 nucleotides of DNA, with no examination of intra-specific variation, and even omitting the orangutan from the analysis

(Ruvolo et al. 1991). Significantly, this was also based on mitochondrial DNA, which has a significant liability in the reconstruction of specific biological history. MtDNA is contributed by the egg at fertilization, such that a child is a clone of mother and unrelated to father, in contrast to the equal relationship of the child to both parents in nuclear Mendelian genes. Imagine a medieval village invaded by Crusaders who make — shall we say — unsolicited contributions to the gene pool. The offspring would be half-Crusader, but that would not be detectable by a mtDNA analysis. There would thus be a significant discordance between biological relationships discernible by mitochondrial and by ordinary nuclear DNA. Indeed, subsequent analyses have found this genetic region to be unreliable in preserving biological history faithfully (Adkins & Honeycutt 1994; Honeycutt et al. 1995).

Quite possibly, then, the scientists who authored the study of the 692 mtDNA bases were using the word "resolution" in a new and heretofore unknown sense.

So (1) it is not at all clear that there are in fact two chronologically distinct speciation events to be resolved here; and (2) if there are, it is not clear that we would be able to tell with the molecular approach in common use now. In a different area of molecular anthropology the sampled polymorphism really is the central focus, namely the data on mitochondrial DNA as it might relate to human origins.

The New Molecular Anthropology: Mitochondrial Eve

The original Cann, Stoneking, & Wilson paper (1987) was a landmark of molecular anthropology, but for a more subtle reason than is generally appreciated. That work was molecular in nature, but didn't cast itself as being "versus morphology." Rather, it contrasted two hypotheses from the literature of morphology or paleoanthropology, subjected them to a test, and sided with one, the "Out of Africa" model. In other words, the researchers self-consciously were using their "molecular" data to try to distinguish between alternatives posed within the "morphology" literature. So the battle lines here are not molecules versus morphology, but molecules plus morphology versus morphology.

The problem with the study is that the authors carried out a fairly standard analysis. That is, they threw their data into a computer, cranked it through a program called PAUP, took the output, and ran with it. It is now clear that the output, the tree they ran with, is sub-optimal (Vigilant et al. 1991; Maddison 1991; Templeton 1992, 1993; Hedges et al. 1992). Nevertheless the data and subsequent studies of genetic diversity show a consistent pattern. For example, the study by Merriwether et al. (1991) broke a human sample up into continental groups, and then asked the computer just to find the breadth of diversity within each group (Figure 6a). They do not find Asia, Africa, and Europe to contain roughly equal amounts of genetic diversity; rather, they find the African sample to contain far more diverse variants than either the European or the Asian. This in turn suggests that the genetic diversity being sampled has been accumulating in Africa longer than elsewhere, as the geographically pooled data imply (Figure 6b).

Cann et al. (1987) calculated that this diversity had been accumulating for about

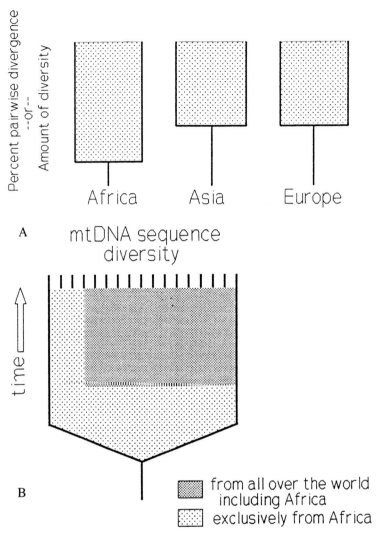

FIGURE 6. (A) General pattern of diversity in human mitochondrial DNA, broken down by
continent (based on Merriwether *et al.* 1991). (B) Pooled data.

200,000 years. Now, how does this relate to the origin of anatomically modem
humans? Unfortunately, not terribly clearly. Mitochondrial DNA can be treated as if
it were a single gene. You can trace the diversity in any gene across the human species
and find the date at which the diversity in that gene originated (Figure 7). Mitochondrial DNA sequences seem to be around 200,000 years old, which may well coincide
with the origin of anatomically modern humans. However, the diversity in histocom-
patibility genes, the ones that encode factors that determine whether skin grafts take,
predates the divergence of humans and chimpanzees, so is on the order of 10 million

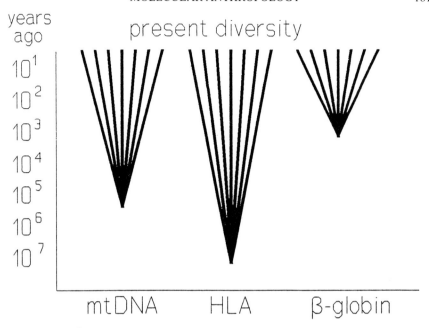

FIGURE 7. Different microevolutionary processes govern the spread of genetic variation. Tracking the time of origin of contemporary human diversity yields different answers for various DNA segments.

years. The breadth of beta-globin diversity is generally taken to be historically tied to agriculture in the Old World — standing water, mosquitoes, malaria, sickle cell anemia — and probably is on the order of thousands of years. So what we have is genetic diversity for different genes, spreading by different microevolutionary processes, having originated at different times. I see the 200,000 year date as a coincidence at best.

On the other hand, the late Allan Wilson, godfather of this research, initiated it with characteristic insight. In 1981, he published a paper actually studying the amounts of intra-specific genetic diversity encountered in the apes, relative to the humans (Ferris *et al.* 1981). Their results were striking: chimpanzees and gorillas had far more genetic diversity detectable in their mitochondrial DNA than humans. Now, since the chimpanzee, gorilla, and human lineages were roughly as old as one another (because they originated around the same time — the trichotomy), there had to be a secondary reason for the human species appearing to be so relatively depauperate in genetic diversity as sampled.

Presumably there was something in the demographic history of our species that caused us to lose the genetic diversity retained by our closest relatives. What might that be? The best way to lose genetic diversity is in founder-effect speciation (Figure 8). If a species originates as a small bud from an ancestral population, then you would anticipate the new species to be considerably more genetically homogeneous than the ancestor.

Breadth of mtDNA diversity
presently detectable

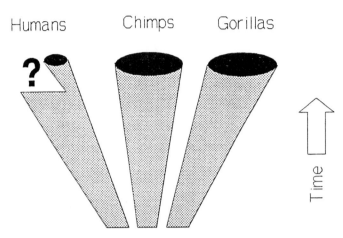

FIGURE 8. Humans, chimpanzees, and gorillas have been accumulating mtDNA diversity for the same amount of time, so a secondary demographic process must be invoked to explain why humans are so much less diverse.

So there are a lot of "ifs" here, but if there really is less genetic diversity in the human species than in chimpanzees or gorillas, and if that is a result of the founder effect operating on the demographic history of our group, and if Africans really are more diverse than Europeans or Asians, then it is entirely possible that we may be sampling the result of a founder effect in Africa marking the origin of our lineage, anatomically modern *Homo sapiens*, a couple of hundred thousand years ago.

Obviously there are other alternatives, but it seems to me that their supporters are preoccupied at the present time with explaining away these data.

The prospects for molecular anthropology lie in its ability to be self-critical. In the phrase "molecular anthropology," "molecular" is just an adjective. It only modifies "anthropology": it helps us tackle old problems with new kinds of data. But these data, while transcending some of the difficulties of traditional datasets, have many others of their own. Because of its mysterious nature, we're obliged to be more critical of molecular anthropology than of traditional anthropology. I think the history of the endeavor bears this out well: without the context of the anthropological framework into which they must fit, molecular data consistently prove to be almost valueless.

Ultimately, if molecular anthropology is not anthropology, what can it ever be? The very phrase "molecular anthropology" itself, after all, was coined by a chemist (Zuckerkandl 1963). I believe that molecular anthropology is a viable and indeed a vital inter-disciplinary field, with a critical role in mediating the domains of genetics on the one hand and anthropology on the other. Molecular anthropology thus can and should serve to elevate genetics on anthropological issues and to elevate anthropology on genetical issues.

On one side, we have *The Bell Curve* (Herrnstein & Murray 1994), invoking genetics para-scientifically to explain group differences in behavior — should not molecular anthropology be crucial to this debate? On the other, we have the Human Genome Diversity Project, proposed externally to anthropology (Cavalli-Sforza *et al.* 1991), which has been represented to revolutionize or even to supplant anthropological knowledge and research (Anonymous 1995a). But without a sound anthropological framework — such as acknowledging the differences between constructed cultural categories and natural biological categories, and the contact and histories of indigenous peoples and their rights (Marks 1995b, 1995c; Anonymous 1995b) — the HGDP has paradoxically begun to raise fundamental doubts about the role of genetics vis-a-vis anthropology.

The issues that confront molecular anthropology today are vexingly similar to those that have always faced those interested in applying genetic data to anthropological issues. In any inter-disciplinary endeavor there are reciprocal intellectual responsibilities incurred, although disappointingly rarely met in molecular anthropology. Our prospects involve continuing to make anthropology better in its integration of molecular research and, more importantly, to make molecular research better as anthropology.

Literature Cited

Adkins, R. M., & R. L. Honeycutt. 1994. Evolution of the primate cytochrome c oxidase II gene. *Jour. Mol. Evol.* 38:215-231.

Anonymous. 1995a. Bias-free interracial comparisons. *Nature* 377:183-184.

_____. 1995b. Genome diversity alarms. *Nature* 377:372.

Bailey, W. J., D. H. A. Fitch, D. A. Tagle, J. Czelusniak, J. L. Slightom, & M. Goodman. 1991. Molecular evolution of the ψ η gene locus: Gibbon phylogeny and the hominoid slowdown. *Mol. Biol. Evol.* 8:155-184.

Billings, P. R. 1992. *DNA on Trial: Genetic Identification and Criminal Justice.* Cold Spring Harbor Laboratory Press, Plainview, New York. 154 pp.

Boyd, W. C. 1940. Critique of methods of classifying mankind. *American Jour. Phys. Anthropol.* 27:333-364.

_____. 1950. *Genetics and the Races of Man.* Little, Brown, Boston. 453 pp.

Brown, W. M., E. M. Prager, A. Wang, & A. C. Wilson. 1982. Mitochondrial DNA sequences of primates: Tempo and mode of evolution. *Jour. Mol. Evol.* 18:225-239.

Caccone, A., & J. R. Powell. 1989. DNA divergence among hominoids. *Evolution* 43:925-942.

Cann, R. L., M. Stoneking, & A. C. Wilson. 1987. Mitochondrial DNA and human evolution. *Nature* 325:31-36.

Cavalli-Sforza, L. L., A. C. Wilson, C. R. Cantor, R. M. Cook-Deegan, & M.-C. King. 1991. Call for a worldwide survey of human genetic diversity: A vanishing opportunity for the human genome project. *Genomics* 11:490-491.

Coca, A. F., & O. Diebert. 1923. A study of the occurrence of the blood groups among the American Indians. *Jour. Immunol.* 8:487-493.

Dangel, A. W., B. J. Baker, A. R. Mendoza, & C. Y. Yu. 1995. Complement component C4 gene intron 9 as a phylogenetic marker for primates: Long terminal repeats of the endogenous retrovirus ERV-K(C4) are a molecular clock of evolution. *Immunogenetics* 42:41-52.

Djian, P., & H. Green. 1990. Vectorial expansion of the involucrin gene and the relatedness of the hominoids. *Proc. Nat. Acad. Sci. USA* 86:8447-8451.

Dulbecco, R. 1986. A turning point in cancer research: Sequencing the human genome. *Science* 231:1055-1056.

Ellis, N., P. Yen, K. Neiswanger, L. J. Shapiro, & P. N. Goodfellow. 1990. Evolution of the pseudoautosomal boundary in Old World monkeys and great apes. *Cell* 63:977-986.

Ferris, S. D., A. C. Wilson, & W. M. Brown. 1980. Evolutionary tree for apes and humans based on cleavage maps of mitochondrial DNA. *Proc. Nat. Acad. Sci. USA* 78:2432-2436.

Ferris, S. D., W. M. Brown, W. S. Davidson, & A. C. Wilson. 1981. Extensive polymorphism in the mitochondrial DNA of apes. *Proc. Nat. Acad. Sci. USA* 78:6319-6323.

Goodman, M. 1962. Immunochemistry of the primates and primate evolution. *Ann. N. Y. Acad. Sci.* 102:219-234.

Green, H., & P. Djian. 1995. The involucrin gene and hominoid relationships. *American Jour. Phys. Anthropol.* 98:213-216.

Hedges, S. B., S. Kumar, K. Tamura, & M. Stoneking. 1992. Human origins and analysis of mitochondrial DNA sequences. *Science* 255:737-739.

Herrnstein, R., & C. Murray. 1994. *The Bell Curve: Intelligence and Class Structure in American Life*. Basic Books, New York. 845 pp.

Hirschfeld, L., & H. Hirschfeld. 1919. Serological differences between the blood of different races. *The Lancet* (October 18):675-679.

Hixson, J. E., & W. M. Brown. 1986. A comparison of the small ribosomal RNA genes from the mitochondrial DNA of the great apes and humans: sequence, structure, evolution, and phylogenetic implications. *Mol. Biol. Evol.* 3:1-18.

Honeycutt, R. L., M. A. Nedbal, R. M. Adkins, & L. L. Janacek. 1995. Mammalian mitochondrial DNA evolution: A comparison of the cytochrome b and cytochrome c oxidase II genes. *Jour. Mol. Evol.* 40:260-272.

Hooton, E. A. 1931. *Up from the Ape*. Macmillan, New York. 626 pp.

Horai, S., Y. Satta, K. Hayasaka, R. Kondo, T. Inoue, T. Ishida, S. Hayashi, & N. Takahata. 1992. Man's place in Hominoidea revealed by mitochondrial DNA genealogy. *Jour. Mol. Evol.* 34:32-43.

Kawamura, A., H. Tanabe, Y. Watanabe, K. Kurosaki, N. Saitou, & S. Ueda. 1991. Evolutionary rate of immunoglobulin alpha noncoding region is greater in hominoids than in Old World monkeys. *Mol. Biol. Evol.* 8:743-752.

Koshland, D. E. 1989. Sequences and consequences of the human genome. *Science* 246:189.

Lewin, R. 1987. *Bones of Contention*. Simon & Schuster, New York. 348 pp.

Livak, K. J., J. Rogers, & J. B. Lichter. 1995. Variability of dopamine D4 receptor (DRD4) gene sequence within and among nonhuman primate species. *Proc. Nat. Acad. Sci. USA* 92:427-431.

Lowenstein, J. 1993. Honestly, those anthropologists. *Pacific Discovery* 46(4):52-53.

Maddison, D. R. 1991. African origin of human mitochondrial DNA reexamined. *Syst. Zool.* 40:355-363.

Manoilov[ff], E. O. 1927. Discernment of human races by blood: Particularly of Russians from Jews. *American Jour. Phys. Anthropol.* 10:11-21.

Manoilov, E. O. 1929. Chemical reaction of blood for definition of sex in man, animals, and dioecious plants. *American Jour. Phys. Anthropol.* 13:29-68.

Marks, J. 1991. What's old and new in molecular phylogenetics. *American Jour. Phys. Anthropol.* 85:207-219.

_____. 1992. Genetic relationships among the apes and humans. *Curr. Top. Genet. Dev.* 2:883-889.

_____. 1993a. Hominoid heterochromatin: Terminal C-bands as a complex genetic character linking chimps and gorillas. *American Jour. Phys. Anthropol.* 90:237-246.

_____. 1993b. Historiography of eugenics. *American Jour. Hum. Genet.* 52:650-652.

_____. 1995a. Learning to live with a trichotomy. *American Jour. Phys. Anthropol.* 98:211-232.

_____. 1995b. *Human Biodiversity: Genes, Race and History.* Aldine de Gruyter, Hawthorne, New York. 321 pp.

_____. 1995c. Anthropology and race. *Nature* 377:570.

Merriwether, D. A., A. G. Clark, S. W. Ballinger, T. G. Schurr, H. Soodyall, T. Jenkins, S. T. Sherry, & D. C. Wallace. 1991. The structure of human mitochondrial DNA variation. *Jour. Mol. Evol.* 33:543-555.

Meyer, E., P. Wiegand, S. P. Rand, D. Kuhlmann, M. Brack, & B. Brinkmann. 1995. Microsatellite polymorphisms reveal phylogenetic relationships in primates. *Jour. Mol. Evol.* 41:10-14.

Miyamoto, M. M., & M. Goodman. 1990. DNA systematics and the evolution of primates. *Annu. Rev. Ecol. Syst.* 21:197-220.

Mourant, A. E., A. C. Kopec, & K. Domaniewska-Sobczak. 1976. *The Distribution of Human Blood Groups and Other Polymorphisms*, 2d ed. Oxford University Press, New York. 1055 pp.

Nelkin, D., & M. S. Lindee. 1995. *The DNA Mystique: The Gene as Cultural Icon.* W. H. Freeman, New York. 276 pp.

Olson, M. V. 1993. The human genome project. *Proc. Nat. Acad. Sci. USA* 90:4338-4344.

Ottenberg, R. 1925. A classification of human races based on geographic distribution of the blood groups. *Jour. American Med. Assoc.* 84:1393-1395.

Pamilo, P., & M. Nei. 1988. Relationships between gene trees and species trees. *Mol. Biol. Evol.* 5:568-583.

Poliakowa, A. T. 1927. Manoiloff's 'race' reaction and its application to the determination of paternity. *American Jour. Phys. Anthropol.* 10:23-29.

Rogers, J. 1993. The phylogenetic relationships among *Homo, Pan* and *Gorilla*: A population genetics perspective. *Jour. Hum. Evol.* 25:201-215.

_____, & A. G. Comuzzie. 1995. When is ancient polymorphism a potential problem for molecular phylogenetics? *American Jour. Phys. Anthropol.* 98:216-218.

Ruano, G., J. Rogers, A. C. Ferguson-Smith, & K. K. Kidd. 1992. DNA sequence polymorphism within hominoid species exceeds the number of phylogenetically informative characters for a HOX2 locus. *Mol. Biol. Evol.* 9:575-586.

Ruvolo, M. 1994. Molecular evolutionary processes and conflicting gene trees: The hominoid case. *American Jour. Phys. Anthropol.* 94:89-113.

_____. 1995. Seeing the forest through the trees: Replies to Marks; Green and Djian; Rogers and Comuzzie. *American Jour. Phys. Anthropol.* 98:218-232.

_____, T. R. Disotell, M. W. Allard, W. M. Brown, & R. L. Honeycutt. 1991. Resolution of the African hominoid trichotomy by use of a mitochondrial gene sequence. *Proc. Nat. Acad. Sci. USA* 88:1570-1574.

Sarich, V. M., & A. C. Wilson. 1967. Immunological time scale for hominid evolution. *Science* 158:1200-1203.

Schull, A. F. 1931. *Heredity.* McGraw-Hill, New York. 345 pp.

Sibley, C., J. Comstock, & J. Ahlquist. 1990. DNA hybridization evidence of hominoid phylogeny: A reanalysis of the data. *Jour. Mol. Evol.* 30:202-236.

Sinnott, E. W., & L. C. Dunn. 1932. *Principles of Genetics*. McGraw-Hill, New York. 441 pp.

Snyder, L. H. 1926. Human blood groups: Their inheritance and significance. *American Jour. Phys. Anthropol*. 9:233-263.

_____. 1930. The "laws" of serologic race-classification studies in human inheritance IV. *Hum. Biol*. 2:128-233.

Templeton, A. R. 1992. Human origins and analysis of mitochondrial DNA sequences. *Science* 255:737.

_____. 1993. The "Eve" hypothesis: A genetic critique and reanalysis. *American Anthropol*. 95:51-72.

Ueda, S., O. Takenaka, & T. Honjo. 1985. A truncated immunoglobulin e pseudogene is found in gorilla and man but not in chimpanzee. *Proc. Nat. Acad. Sci. USA* 82:3712-3715.

Ueda, S., Y. Watanabe, N. Saitou, K. Omoto, H. Hayashida, T. Miyata, H. Hisajima, & T. Honjo. 1989. Nucleotide sequences of immunoglobulin-epsilon pseudogenes in man and apes and their phylogenetic relationships. *Jour. Mol. Biol*. 205:85-90.

Vigilant, L., M. Stoneking, H. Harpending, K. Hawkes, & A. C. Wilson. 1991. African populations and the evolution of human mitochondrial DNA. *Science* 253:1503-1507.

Watson, J. D. 1990. The human genome project: Past, present, and future. *Science* 248:44-51.

Zuckerkandl, E. 1963. Perspectives in molecular anthropology. Pages 243-272 *in* S. L. Washburn, ed., *Classification and Human Evolution*. Aldine, Chicago, Illinois.

Index

Index

A

ABO allele frequencies 170
adaptive radiation 10, 74, 97, 103, 150
Africa 3, 6-7, 11, 14-15, 21, 23, 26-28, 32-33,
36-37, 76, 78, 81, 87-88, 93, 100-104, 107,
112, 115, 119-121, 125-126, 130-133, 135,
143-144, 147-150, 152-159, 161-163, 165,
170, 179, 182; See also Chesowanja; See
also East Africa; See Florisbad; See also
Irhoud; See also Kabwe; See Koobi Fora;
See also KRM (Klasies River Mouth); See
also Kromdraai; See also Makapansgat;
See also Nariokotome; See also Olduvai;
See also Olorgesailie; See also Omo; See
also Peninj; See also South Africa; See also
Sterkfontein; See also sub-Sahara; See also
Swartkrans; See also Taung; See also Tighenif; See also Turkana
African origins (models of; recency of) 23-24,
26-28, 31, 35, 184
AHRM; See African origins (models of; recency
of)
Allen's Rule 149
Alu insertion 22
anagenesis 22, 34, 59
anterior dentition; See dentition
anthropophagy 16
AOAM; See African origins (models of; recency
of)
apes; Africa 5, 106, 135
apomorphy 16-20, 24, 62, 65-66, 70, 94, 116-
117, 121, 124-125, 127, 129-130
Arago (Spain) 16-17, 121, 125
archaic *Homo sapiens* 121, 126, 133, 136, 151,
159
Ardipithecus 3, 80; *ramidus* 3, 80, 82, 89
Asia; See Dali (China); See Hathnora (India);
See India; See Indonesia; See Jinniushan
(China); See Maba (China); See Narmada
Valley (India); See Yunxian (China); See
Zhoukoudian (China)
Asian Neanderthals 127-128
Atapuerca (Spain) 9, 16-17, 118, 125-126, 131,
133, 139, 145, 152-153, 155; Gran Dolina
16; Sima 9, 16, 131, 152
Aurignacian assemblages 149-150, 152-153, 165
Australia 12, 118, 120, 147-149, 157, 160, 170
australopithecine; gracile 78, 99, 109; robust
71, 77-81, 84, 86-88, 99, 102, 104, 142; taxa
14, 110
Australopithecus 5-6, 9, 50, 78, 80-82, 87-89,
93-104, 106, 110-111, 113, 135-136, 138,
142-143, 157-160, 162; *aethiopicus* 78, 80-
87, 92-93, 96; *africanus* 5, 78-84, 86-89, 92,
96-97, 99-100, 104-105, 110-111, 160; *anamensis* 80, 82; *boisei* 78, 80-84, 86, 88-89,
96, 102; *robustus* 78, 80-84, 86, 92, 96; See
also *Homo africanus*; See also *Paranthropus*; See also *Paranthropus robustus*

B

basicranial flexion 82, 84, 86, 146
biogeography 11, 76, 161
biotic subregions; lowland tropical forest 28;
Saharan 11, 28; Sahelian 28; sa-
vanna/grasslands 28
bipedality 80, 97, 105-106, 110-111, 113, 137-
138, 141, 159-161, 163, 165
birth interval; See demographics
Black Skull (WT 15000) 78, 84
blood groups 2, 21, 169-171, 183, 185-186
bone chemistry; See isotopes and strontium
bottlenecks 26, 31, 38
brain size; See endocranial capacity
brain/body-size ratio 141
Broca's area 142-143
Broom, Robert 5, 87, 93-96, 101

C

carnivores 138, 140-141, 146-147
caves 15, 132, 141, 144-145, 154, 160
cheek teeth; See dentition
Chesowanja (Africa) 94, 144
chimpanzee 5, 21, 36-37, 80, 82, 155, 172-174,
177, 180-182, 186; See also Pan
China; See Dali; See Jinniushan; See Maba;
See Yunxian; See Zhoukoudian
chronocline 62-66, 69
chronometric dating 4
cladistic analysis 4-5, 12, 36-37, 51, 55-56, 62,
66, 69-71, 73, 75-76, 79-80, 105, 111, 117,
121, 135, 155
cladistics 38, 55, 73-75
cladogenesis 10, 59, 118, 124
cold-adaptation 124
cosmetics; mineral pigments 146; See also
ornaments
cranial vault; See cranium, vault
craniometrics 12, 36
cranium 14-16, 18, 33, 81, 87, 94, 97, 99-100,
102, 106-107, 110, 112, 119, 131, 147, 155;
vault 17-18, 20, 33, 78, 82, 92, 107, 123, 127,
145; See also neurocranium
Cro-Magnons 147, 152, 160, 164
Cuvier, Baron Georges 56-57, 73

D

Dali (China) 18-19, 126, 129
Dart, Raymond 5, 94, 102
demes 8-9, 12-15, 17-22, 24, 27-29
demographics; adult female longevity 24; birth interval 24, 149; doubling time (population) 25; growth rates 24-25, 34; life expectancy 24; parameters 24, 28; population estimates 26; population expansion 25-26, 36; population sizes 27
dentition; anterior 82, 84, 86, 92, 97-98, 106; cheek teeth 78-82, 85-86, 93, 95, 97-101; See also microwear studies
Depéret, Charles 59, 73
dietary adaptations 3, 93
dispersal of *Homo*; general 1, 6-7, 10-11, 26-27, 32, 38, 133, 147
diurnal activity 143
Dmanisi (Georgian Caucasus) 6, 10, 15, 27, 34, 145, 156
DNA 3, 21-22, 25-26, 29-30, 33-38, 167-168, 172-186; fingerprinting 168; hybridization 172, 185; See also microsatellite; See also mtDNA; See also nDNA

E

East Africa; rift valley and localities 10
East African Rift 50
elephants 59, 67, 69, 75, 139, 145
enamel 98-99, 110, 138, 142-144, 146, 150, 159-160, 162; hypoplasia 138, 144, 146, 160, 162; thickness 138, 159
encephalization 82, 84, 86, 159
endocranial capacity 80-83, 131; brain size 9, 79-80, 82, 105-106, 108-111, 113, 136, 144
England; Swanscombe 16, 121, 126
erectus, Homo; See *Homo erectus*
ergaster, Homo; See *Homo ergaster*
Erq el Ahmar (Jordan Valley) 6, 27
Eurasia 1, 3, 6, 15, 17, 27-28, 32, 118, 126, 130, 158, 160
Europe 7, 11-12, 16-17, 23, 27, 33, 36, 48, 53-54, 119-121, 126, 130-133, 144-152, 155, 161-162, 170, 179; See Arago (Spain); See Atapuerca (Spain); See Mauer (Germany); See Petralona (Greece)
European Neanderthals 122, 127-128

F

female reproductive span; See demographics
fire, use of 144, 147, 150, 154, 157
Florisbad 15, 125-126, 131
food procurement 138; fishing 147-148; hunting 34, 138, 140, 143, 145, 147-151, 162; trapping 147

fossil record (general) 4-9, 11-13, 17, 21-22, 29, 32, 38, 47, 49-51, 53, 56, 63-64, 66-67, 69-72, 74-75, 86, 106, 112, 116-117, 141
Founder effect 176
functional complexes 82-84

G

gene flow 21, 23, 26, 29, 32, 125, 128, 177
genetic diversity 21, 28, 178-179, 181-183
genetic drift 32
geochronology 28, 30, 159
Glagahomba; See Sangiran (Indonesia)
gnathic remains 11; See also jaws
Gorilla 5, 12, 21, 81-82, 138, 145, 172-174, 177, 181-182, 185-186; See also apes, African
Greece; See Petralona

H

habilis, Homo; See *Homo habilis*
habitation; See caves; See rock-shelters
Haeckel, Ernst 57, 74
Hathnora (India) 18-19
heidelbergensis, Homo; See *Homo heidelbergensis*
helmei, Homo; See *Homo helmei*
hemoglobin 177; genes 177
Hennig, Willi 55, 62, 66, 68, 72, 74, 78, 88
Hominidae 5, 9, 33, 88, 94, 102, 112, 162; Homininae 5; Hominini 5, 48, 52
hominin evolution; orthogeneticism 10; progressivism 10
hominins 1, 6-12, 15-24, 27-29, 31-32, 47, 51, 53, 98, 112, 114; European 11; fossil record 7-9, 21, 29, 32
Hominoidea 30, 32, 79, 184
Homo 1, 5-6, 9-10, 13, 20, 22, 24, 27, 30-33, 35, 37-39, 47-54, 71-72, 75, 77-84, 86-87, 89, 92-94, 96-97, 99, 101-102, 104-119, 121-126, 129-133, 135-136, 138, 141, 143-146, 148, 150-151, 153, 155-166, 169, 182, 185; *erectus* 6, 9-11, 19, 28, 33, 35, 51, 54, 71, 96, 105-108, 111-112, 115, 117-119, 121-124, 126, 130-133, 135-136, 138, 143-146, 151, 153, 155, 160-162, 164-166; *ergaster* 6-7, 10-11, 14, 27, 96, 105, 107-108, 111, 118, 121, 143, 151; *habilis* 6-7, 13, 71, 80, 82, 89, 96-97, 105-113, 115, 130, 135-136, 141-143, 151, 160, 163; *heidelbergensis* 16, 96, 115, 117, 124-126, 130, 144; *helmei* 15, 115, 126, 130; *leakeyi* 14, 121; *neanderthalensis* 17, 48, 96, 111, 115, 117, 124-126, 130, 132, 136, 144, 146, 150, 160; *rhodesiensis* 14, 115, 126, 130; *rudolfensis* 6, 13, 96, 105, 108, 110-111, 151; *sapiens* 1, 5, 9, 13, 17, 22, 29-30, 32-33, 35, 39, 47-51, 53, 71-72, 96, 106, 108, 111, 115, 117-119, 121, 123-126,

129-133, 135-136, 144-147, 150-151, 154-155, 157, 159-160, 162-163, 169, 182; *sapiens neanderthalensis* 48; *sapiens palestinus* 9, 17
homology 23, 52
homoplasy 11, 52-53, 63, 66, 70, 77-78, 80, 83, 86, 89, 111, 128
horses 69, 74
human evolution 2-4, 7, 10, 20, 29, 33-38, 51, 53-54, 72, 79, 101-102, 118-119, 129, 132-133, 135-136, 141, 143, 157, 159, 161, 168, 183
human genome 168, 183-186
Human Genome Diversity Project 183
hunter/gatherer 28, 137, 141
hunting, techniques of 145, 149
hyaenas 152, 163
hypoplasia; See enamel

I

immunoglobulin 174, 184, 186; genes 174
India; See Hathnora; See Narmada Valley
Indonesia 11, 37, 113, 118-119, 163; Java 6-7, 10, 19, 37, 113, 126, 130, 163; Java, Perning 6-7, 10; Java, Sangiran 6-7, 10-11, 19-20, 35, 118; See Ngandong (Java); See Sangiran (Java)
involucrin (protein) 174, 177, 184
Irhoud 15, 17, 125-126, 129, 147
isotopes; carbon-13 139; nitrogen 139, 156; oxygen 139

J

Java (Indonesia); See Indonesia, Java
jaws 78, 81-82, 145, 147, 152; See also gnathic remains
Jinniushan (China) 18-19, 126
Johnson Immigration Restriction Act of 1924 168
Jordan Valley 163; See also Erq el Ahmar and Ubeidiya

K

Kabwe 14-15, 104, 125
KNM-ER 3733 115
Koobi Fora (East Turkana, Africa) 39, 89, 107-108, 113-115, 144, 157, 166
KRM (Klasies River Mouth) 15
Kromdraai (South Africa) 87-88, 93-94, 96, 100-101, 104

L

Lamarck, Jean Baptiste 55-57, 73-74
leakeyi, Homo; See *Homo leakeyi*

Levant; See Qafzeh; See Skhul
Levantine Neanderthals 127
life expectancy 24
limb proportions 128, 133, 137, 143, 164
locomotion 106, 111
lowland tropical forest 28; See also biotic subregions

M

Maba (China) 18-19, 126, 129
macroevolution 174
Makapansgat (South Africa) 100, 102, 104, 159
malaria, resistance to 178, 181
masticatory apparatus 98
matrix correlation 23
Mauer (Germany) 16-17, 125
metaspecies 70
microevolution 174, 181
microsatellite (DNA) 22, 34
microwear studies 98, 100, 102-103
Middle East; See Israel; See Jordan Valley; See Levant
Miocene 66-67, 69, 74-75, 124, 174
Mitochondrial Eve; See African origins (models of; recency of)
molecular anthropology 173-174, 179, 182-183, 186
molecular biology 2-3, 30
molecular clock 21, 184
molecular systematics 172
morphoclines 51, 62-66, 69, 82-83
morphological distance 23
mosaic evolution 33, 65, 154
mosaic morphology 15
MRE (multiregional evolution); See multiregional origin
mtDNA 21-22, 25-26, 29-30, 36, 179, 182
multiregional evolution (MRE); See multiregional origin
multiregional origin 23; MRE (multiregional evolution) 31-32; Regional Continuity (model) 27, 31
multivariate analyses 12, 21, 33, 35, 37-38, 132

N

Nariokotome 14, 143, 153, 161-162, 165
Narmada Valley (India) 19, 126
nDNA 21-22, 30
neanderthalensis, Homo; See *Homo neanderthalensis*
Neanderthals 9, 16-17, 23, 48, 53-54, 72, 115-116, 118, 122-133, 138-139, 144-150, 154-160, 162-164; See also Asian, European, Levantine; See also *Homo neanderthalensis*
Near East; See Middle East
neurocranium 110

Ngandong (Indonesia (Java)) 9, 20, 118, 130

O

Olduvai; LLK-II 14-15
Olduvai (basin and gorge) 6-7, 10, 14-15, 27, 87-89, 94, 99-100, 102-103, 107-108, 113, 115, 121, 142-143, 151, 155, 158, 160, 163
Olorgesailie (Africa) 143, 160
Omo 33, 81, 87, 94, 101-102, 107, 112, 130, 147, 155; Shungura Formation 81, 107, 155, 159
orangutan 5, 172-173, 178; See also *Pongo*
ornaments 141, 146, 148; See also cosmetics
orthogeneticism; See hominin evolution , orthogeneticism
orthognathism 82, 84, 86

P

p-demes; See paleodemes
pair-bonding 138
paleoanthropology 2-4, 6, 30, 47, 49-53, 88, 132, 157, 168, 179
paleodemes 1, 9-10, 13-22, 29
paleoenvironment 4, 25, 30, 158
Paleolithic; Middle 27; Upper 27
Paleolithic 6, 25, 27, 38, 115-116, 122, 126-128, 146, 149-150, 152-153, 156, 158, 161-163, 165
paleontology; human 9-10, 20, 30, 38, 54, 133, 163
palestinus, Homo sapiens; See *Homo sapiens palestinus*
Pan 5, 22, 36, 82, 185; See also apes, African
parallel evolution 77, 84, 86
Paranthropus 5-6, 9, 14, 80, 87, 93-102, 106, 110-111, 163; *boisei* 93, 96, 100; *crassidens* 14, 80, 87, 93, 96, 100, 102; *robustus* 87, 93, 96, 100, 163; See also *Australopithecus boisei*; See also *Australopithecus robustus*
paraphyly 10, 63, 69-70, 176
Paraustralopithecus aethiopicus 87; See also *Australopithecus aethiopicus*
parsimony 52, 66, 69, 71-72, 74-75, 82-83, 89
PAUP 52, 72, 75, 179
Peninj (Africa) 94
Penrose analysis 122
Petralona (Greece) 16-17, 121, 125, 145
phyletic gradualism 51, 59, 73
Pleistocene 3, 6, 9, 11, 16, 19, 22-29, 31-37, 53, 87, 89, 101-102, 104, 111-112, 115-116, 118-122, 124-126, 130-133, 151-157, 159-165
plesiomorphy 16-18, 62, 117, 121, 125-129
Pliocene 3, 6, 28, 88, 102, 117, 126, 157, 161
polymorphisms 22, 31, 34, 36, 116, 169-170, 174-175, 178-179, 184-185
polytypic species 31

Pongidae 5; Dryopithecinae 5; Ponginae 5; Pongini 5; Sivapithecini 5
Pongo 5, 81, 98
population genetics 2, 29-30, 32, 37-38, 174, 185
populations; See demographics
posture 106, 111
predation 138, 157
progressivism; See hominin evolution, progressivism
punctuated equilibrium 51

Q

Qafzeh (Levant) 9, 17, 72, 115-116, 122, 126-130, 132-133, 146-148, 160-161, 164
Qafzeh 6 127-128, 130

R

radiometric dating 127, 155
Ramapithecus 104, 172
RAO; See African origins (models of
Regional Continuity (model); See multiregional origin
Replacement models; See African origins (models of; recency of)
rhodesiensis, Homo; See *Homo rhodesiensis*
Robinson, John T. 87-88, 93-100, 102-103, 106, 111, 113
rock-shelters 145, 152
rudolfensis, Homo; See *Homo rudolfensis*
Russia 148, 162, 171, 184

S

Sangiran (Indonesia (Java)); Brangkal 19; Glagahomba 19; Trinil 19-20
sapiens, Homo; See *Homo sapiens*
savanna/grasslands 28; See also biotic subregions
scavenging 140-141, 146
seasonality (of activities) 141, 147, 158
sexual dimorphism 81, 88, 101, 119, 137, 142, 145-146, 156, 159, 161
Shungura Formation; See Omo
Siberia 17, 147-148, 150, 156
sickle cell anemia 181
Simpson, George Gaylord 8, 12-13, 20, 37, 49, 54, 94, 103, 105, 113, 162
single-species hypothesis 20, 50, 113
SK 847 102, 107
Skhul (Levant) 9, 17, 72, 115-116, 122, 126-131, 133
Skhul 4 128-129
Skhul 5 127-129
Skhul 9 128-129
social networks 146, 148-150

South Africa 15, 78, 81, 87-88, 93, 95-96, 101-102, 104, 112, 144, 147-148, 153-156, 158, 162
Southwestern Asia; See Middle East
species concepts 8, 11, 33, 38, 50, 73, 116-117
Sterkfontein (South Africa) 9, 87, 93, 96-97, 100, 102, 104, 107
stone transport 143
stratocladistics 70, 74
stratophenetics 59
strontium 139, 161
Sts 71 87, 96
sub-Sahara 11
Sundaland 7, 11, 19
Swanscombe, England; See England, Swanscombe
Swartkrans 9, 14, 87, 94, 96, 100, 102, 104, 107, 112, 144, 154, 158, 162-163
synapomorphy 17, 24, 65-66, 116, 124-125, 129-130

T

taphonomy 4, 30, 104, 154-155, 160, 163
Taung (South Africa) 100, 104
Ternifine; See Tighenif
Tighenif 14
tools; cores 55; flakes 98, 142, 144; hammerstones 143; handaxe 102, 143-145; Oldowan 143, 151, 155; prepared-core technology 145; projectiles 139, 141, 143, 147-148, 150; scrapers 141, 149, 152; spears 145, 147, 160; spheroids 143-144, 161
Trinil (Indonesia (Java)); See Sangiran (Java)
Turkana 6, 10, 33, 88-89, 94, 100, 102, 112, 157, 161; East 39, 89, 107-108, 113-115, 144, 157, 166; West 33, 100

U

Ubeidiya (Jordan Valley) 6-7, 163

V

venous drainage; pattern of 142
vocal tract 138, 144

Y

Y chromosome 22, 34, 36, 38, 174
Yunxian (China) 17-18, 132

Z

Zhoukoudian (China) 18-19, 118, 154, 166
Zinjanthropus boisei 87; See also *Australopithecus boisei*